EMERGING TECHNOLOGIES

IN DISTANCE EDUCATION

> **Edited by** *George Veletsianos*

Issues in Distance Education

Series Editor: Terry Anderson

Distance education is the fastest-growing mode of formal and informal teaching, training and learning. Its many variants include e-learning, mobile learning, and immersive learning environments. The series presents recent research results and offers informative and accessible overviews, analyses, and explorations of current issues and the technologies and services used in distance education. Each volume focuses on critical questions and emerging trends, while also taking note of the evolutionary history and roots of this specialized mode of education and training. The series is aimed at a wide group of readers including distance education teachers, trainers, administrators, researchers, and students.

Series Titles

© 2010 George Veletsianos

Published by AU Press, Athabasca University
1200, 10011 – 109 Street
Edmonton, AB T5J 3S8

Library and Archives Canada Cataloguing in Publication

Emerging technologies in distance education / edited by George Veletsianos.

(Issues in distance education, ISSN 1919-4382)
Includes bibliographical references and index.
Also available in electronic format (978-1-897425-77-0).
ISBN 978-1-897425-76-3

1. Distance education–Technological innovations.
2. Educational technology.
I. Veletsianos, George
II. Series: Issues in distance education series (Print)

LC5800.E54 2010   371.3'58   C2010-903417-1

Cover and book design by Natalie Olsen, kisscutdesign.com.
Printed and bound in Canada by Marquis Book Printing.

Please contact AU Press, Athabasca University at aupress@athabascau.ca for permission beyond the usage outlined in the Creative Commons license.

A volume in the Issues in Distance Education series:
1919-4382 (Print)
1919-4390 (Online)

GEORGE VELETSIANOS

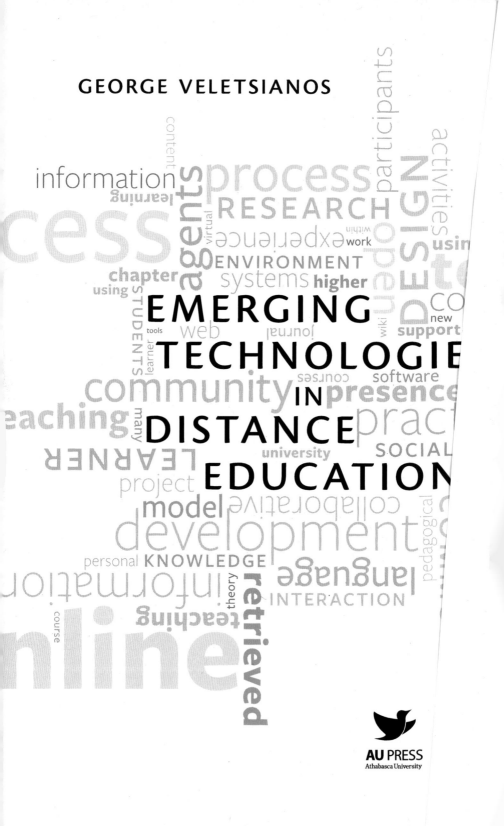

EMERGING
TECHNOLOGIE
IN
DISTANCE
EDUCATION

AU PRESS
Athabasca University

Editor Acknowledgements

First, I would like to thank the chapter authors for contributing their valuable insights to this collection and for being willing to participate in this project.

I would also like to thank Athabasca University Press for their assistance, dedication, hard work, and support.

Special thanks also go to Terry Anderson, the series editor, who provided valuable guidance and assistance whenever I asked.

Finally, to all those individuals whose work and writings influenced my thinking, thank you for sharing your knowledge.

Part 3

Social, Organizational, and Contextual Factors in Emerging Technologies Implementations

Part 4

Learner-Learner, Learner-Content, and Learner-Instructor Interaction and Communication with Emerging Technologies

INTRODUCTION

> *George Veletsianos*

Emerging technologies have been heralded as providing the opportunities and affordances to transform education, learning, and teaching. Nevertheless, scholarship on the opportunities of emerging technologies in the context of online *distance* education has been minimal. Most often, researchers, designers, and educators present a description of how such technologies can be used in face-to-face and hybrid courses, but not in distance education courses. Additionally, distance education researchers and practitioners reside in varied academic domains, rendering the sharing and dissemination of their work a formidable task. As a result, the picture of how such technologies are used in distance education is fuzzy. In this book, therefore, we sought to amalgamate work in the use of emerging technologies to conceptualize, design, enhance, and foster distance education. This edited volume intends to harness international experiences, dispersed knowledge, and multidisciplinary perspectives for use by both members of research communities and innovative distance education practitioners. Notably, contributors from eight countries (Australia, Canada, Cyprus, Finland, Greece, Israel, the United Kingdom, and the United States) discuss a broad range of issues. Whether training teachers and designers in Canada, promoting the use of wikis within a single institution in Israel, or engaging teachers and students in worldwide climate change dialogue, the thread connecting these chapters is the use of emerging technologies in distance education.

The book begins by discussing the foundations and meaning of *emerging technologies*. George Veletsianos (chapter 1) notes that the term "emerging technologies" is often used haphazardly without a clear understanding of what it really means. The conceptualization of the term proposed in chapter 1 situates the chapters that follow and establishes a common ground upon which future conversations can be extended. Terry Anderson (chapter 2) solidifies the foundations of this book by reviewing established and contemporary learning and instructional theories intended to guide the utilization of emerging technologies in distance education. Importantly, the work presented in later chapters of this volume can be traced back to the theoretical foundations discussed by Anderson. In turn, Wellburn and Eib (chapter 3) investigate the opportunities and complexities afforded by emerging technologies and ask readers to explore the meaning of our roles as experts, amateurs, authors, learners, educators, and audiences. In the same way that other authors in this volume highlight (a) the power of the method, and (b) the power of the technology to transform and widen the methods we use rather than the medium per se (e.g., chapters 2, 5, 6, 7, 14), Wellburn and Eib ask us to envision how the affordances presented to us by emerging technologies can empower us to change the ways we teach and learn. While "connected and social" distance education is a facet of emerging technologies that is discussed in chapter 3 and investigated throughout the book, Lee and McLoughlin (chapter 4) examine the potential of the participatory nature of the Web to rectify traditional distance education problems and foster improved learning experiences. Central to the arguments and examples presented in this chapter is the idea that emerging technologies can enhance authentic and social learning experiences by enhancing presence, community, interaction, and participation.

The second part of the book focuses on emerging pedagogical approaches that are facilitated by emerging technologies. In chapter 5, Doering, Miller, and Scharber illustrate and exemplify how the ideas presented in the introductory chapter of this volume are evident in practice by focusing on the idea that the way we use technology matters more than the tool we use. Specifically, they introduce Adventure

Learning as a framework that provides students and teachers with the opportunity to engage in real-world experiences while collaborating and interacting with explorers, students, and content experts at various locations throughout the world. Collaboration between learners and learner-educators and the use of a multiplicity of emerging technologies to design engaging learning experiences are also evident in the chapter authored by Couros (chapter 6). In this chapter, Couros reflects upon and discusses the theoretical, pedagogical, technological, and philosophical foundations of a graduate-level educational technology course delivered at a distance. This course was informed by an open teaching model and made extensive use of the author's personal learning network to facilitate learner integration into a persistent online learning community. Perry and Edwards (chapter 7) extend our thinking on learning communities by arguing that online cultures of community are founded on artistic elements. Artistic Pedagogical Technologies, situated within philosophical, theoretical, and pedagogical considerations, are thus presented as teaching strategies intended to enhance presence, community, and interaction. The section on emerging pedagogical approaches concludes with a chapter from Laouris and colleagues (chapter 8), who describe the science of dialogic design and its use within emerging technologies to develop a new methodology for distance-based disciplined and democratic dialogue. Examples in which the dialogic design process was embedded within emerging technologies are also presented.

Next, five chapters investigate the complex social, organizational, and contextual landscape of emerging technology implementations in distance education. First, Martindale and Dowdy (chapter 9) introduce Personal Learning Environments (PLEs) as broad and holistic learning landscapes, as well as specific collections of tools that facilitate learning. These authors explore the implementation, use, adoption, and challenges faced by PLEs (both in terms of personal and institutional adoptions) while positioning PLEs as powerful environments in the quest for informal and self-directed learning. Whitworth and Benson (chapter 10) examine directive and responsive learning management systems and investigate how one specific responsive emerging technology, Moodle, came to be utilized in two institutions with divergent aims, communities, and practices. Importantly, in line with ideas presented in chapters 1 and 11, Whitworth and Benson demonstrate both how the actual technology *influenced* distance education practice, and how the use, implementation, and adoption of the technology *were influenced by* educational practice. Further granulations of this idea are posited by Hagit Meishar-Tal, Yoav Yair, and Edna Tal-Elhasid (chapter 11), who discuss the experience of implementing wikis at the Open University of Israel. The chapter examines technological, pedagogical, and administrative perspectives related to institutional implementation, and extends this discussion into matters relating to wiki diffusion and sustainability. Importantly, the authors highlight one important dimension of emerging technologies: the possibility of the institution adjusting to the emerging technology such that the technology becomes part of the institution's culture of learning and teaching.

One emerging technology that can yield insights into technology adoption, diffusion, and use within an institution is web analytics, which is the focus of the next chapter (chapter 12). Rogers, McEwen, and Pond introduce web analytics as an emerging tool used in the design and evaluation of distance education. Specifically, the authors explain how web analytics can be utilized to gain knowledge about, and insight into, student behaviors, outcomes, and engagement. By learning if, how, when, and to what extent learners engage with web-based courses, instructors and distance education providers can make efficient and effective curricular and pedagogical decisions with regards to distance and web-based courses.

Caladine and colleagues (chapter 13) conclude this section of the book by investigating key issues with regards to employing Internet Protocol Video Communications in distance education. While interest in video communication has been expanding in recent years, Caladine et al. note that distance education instructors and managers lack the knowledge and skills to effectively and efficiently harness video communications. Beyond the importance of video, however, chapter 16 introduces a crucial point for the study and use of emerging technologies:

"A repeating dilemma will arise with each new wave of technology: *Should this be used for formal education or is it a personal/social tool better left in the realm of informal communication?*" Anderson (chapter 2), Wellburn and Eib (chapter 3), Martindale and Dowdy (chapter 9) , and Kop (chapter 14) implicitly raise the same question. While a strong desire (and perhaps pressure) exists to employ new and emerging technologies in formal distance education (see chapter 1), it is important that we critically evaluate (and experiment with) a set of technologies with respect to the opportunities that they afford.

The final section of this book deals with interaction and communication with emerging technologies, a theme that permeates educational technology discussions in general, and this book in particular. Kop (chapter 14) presents a case study of how emerging technologies can be used for true dialogue in the context of an informal and comfortable online place that enables a sense of "nearness" and "presence." In line with Anderson's theoretical foundations (chapter 2), Kop highlights the value of communication and interaction. In addition, Kop introduces ideas expanded upon by other chapters within this volume, including institutional control (chapter 10), empowered instructors (chapter 6), and the distinctions between "amateur" students and "expert" instructors (chapter 3). Wang, Calandra, and Yi (chapter 15) explore cross-cultural technology use in their investigation of the affordances provided by Multi-User Virtual Environments for learning English as a foreign language. One of the most important lessons highlighted by Wang and colleagues (and also discussed in chapters 2, 5, 6, and 16) relates to the fact that interaction is of fundamental importance to the design of successful learning experiences. In particular, the authors note that when engaging learners in language learning within MUVEs, designers and instructors need to consider the possible interactions between learners and (a) their own avatars, (b) the avatars of others, and (c) the virtual environment. Heller and Procter (chapter 16) expand the discussion of interaction within virtual worlds by focusing on virtual characters. Specifically, they review the field of animated pedagogical agents and concentrate on actor agents that are able to participate in pedagogical simulations and activities. Reformulating the discussion of avatars in virtual worlds in the context of pedagogical agents, Heller and Procter highlight interaction and communication between learners and avatars, as well as the narrative in which the learner experience is situated.

I hope that you find this book enjoyable and worthwhile for your practice and research. Personally, I view the work presented here as the beginnings of a larger conversation about education, technology, and universities, rather than the final words of wisdom from academics. I would therefore like to see this work extended through conversations in conferences, journals, and web postings, further refining the ideas presented and further aiding in enhancing research *and* practice. It is only through conversation and refinement of ideas that we can improve education. This book, by being offered freely and openly to anyone interested, aims to do just that.

Part 1
Foundations of Emerging Technologies in Distance Education

A DEFINITION OF EMERGING TECHNOLOGIES FOR EDUCATION

1

> *George Veletsianos*

Acknowledgements

This chapter benefited greatly from the contributions of numerous people, most of whom I have never met. I would like to thank George Siemens for posting my question on searching for a definition of emerging technologies on Twitter and on his blog. I am also deeply thankful to all the individuals who answered that question. I also want to recognize the contributions of a number of authors whose work appears in this edited collection: Elizabeth Wellburn, B.J. Eib, and Alec Couros, who encouraged me to explicitly state the implications of "context" on the proposed definitions; Hagit Meishar-Tal and Andrew Whitworth, who highlighted the sociological aspect of the proposed ideas; and Bob Heller, who encouraged me to think more broadly about these technologies. Thank you all for contributing to this definition.

Abstract

The term "emerging technologies" is often used without a clear meaning or definition. My aim in this chapter is to understand the meaning of the term while at the same time exploring what a clear understanding of emerging technologies means for technology-enhanced learning. Combining previous conceptualizations of the term, I propose that emerging technologies are tools, concepts, innovations, and advancements utilized in diverse educational settings to serve varied education-related purposes. Additionally, I propose that ("new" *and* "old") emerging technologies are evolving organisms that experience hype cycles, while at the same time being potentially disruptive, not yet fully understood, and not yet fully researched. These ideas bring to the surface important issues relating to the use of technology in education.

Introduction

Technological innovation and advancements have brought about massive societal change. In comparison, technology's impact on education, teaching, and learning has been rather limited (Bull, Knezek, Roblyer, Schrum, & Thompson, 2005). While expectations have run high about instructional radio, television, personal computers, computer-based instruction, the Internet, Web 2.0, e-learning, m-learning, the latest technological innovation of our times, and the impact of these tools and technologies, results have often been disappointing (see Cuban, 2001): "showcase" learning environments, disengaged students, and technology-enhanced instruction that merely replicates face-to-face teaching seem to be the norm and the standard to which we have become accustomed, rather than the exception.

As a field that seems to find joy in the development of acronyms, terms, and catchy descriptors (think *i-learning, student 2.0, education 3.0*) we seem to quickly traverse innovations in the hope that the next technological advancement will be our holy grail. The focus of this book, however, is not on all previously used educational acronyms. The focus is on the often-misused, haphazardly defined, ill-applied, and all-encompassing term of "emerging technologies" as used in educational contexts in general, and distance education in particular. Siemens (2008, ¶ 1) makes a similar argument when he states that "terms like 'emergence,' 'adaptive systems,' 'self-organizing systems,' and others are often tossed about with such casualness and authority as to suggest the speaker(s) fully understand what they mean."

If you think that I am being unfair in my description of emerging technologies for education, ask your colleagues at your next conference gathering to describe (or dare I say, define) emerging technologies. The majority of your colleagues will agree that emerging technologies describe *new* tools with *promising potential*. If you feel brave, you might ask what *new* means, but let me warn you that you may find yourself faced with rolling eyes and questioning looks. In my questioning, I was not able to find an adequate definition of the term, or at least a description that differentiates between technologies as emerging or non-emerging (e.g., developed or established). Searching prior literature for a definition is the logical next step. Yet again, you will be quickly disappointed. Not only is the literature plagued with casual mentions of the term, it also spans multiple and divergent fields: educators from multiple academic disciplines employ the services of emerging technologies to pursue academic endeavours. Does one search the literature from all academic disciplines? Or does one focus on his/her own discipline? Do emerging technologies transcend academic foci? Do we just search the distance education and instructional design literature? Or do we examine individual content areas, such as nursing, art, and social science education?

In the sections that follow, I argue that the utilization of emerging technologies for education transcends academic disciplines. After discussing my attempts to locate a clear discussion/understanding of "emerging technologies," I put forth my own definition of the term and conclude with thoughts on the implications of this definition.

Emerging Technologies: An Interdisciplinary Notion

The view espoused in this chapter (and in this collection) is that the term "emerging technologies" transcends academic disciplines and activities, and can be defined independently of its specific application to educational endeavours. While some innovations might be more appropriate for specific content areas than others (e.g., Geometer's Sketchpad for mathematics-related disciplines), and technological affordances may render some tools more appropriate for certain purposes than others (e.g., wikis and blogs for community-focused and writing-intensive modules), on the whole, emerging technologies can be applied to diverse disciplines. A November 2008 search on the PsychInfo database, for example (for papers published from 2000 to 2008 that include the keywords "emerging technologies" and "education"), yielded 255 results. The diversity in these results is clear: emerging technologies are used in nearly every field imaginable, with teacher training, instructional design, language learning, distance education, e-learning, adult education, and medical education prominently appearing on the list. The accepted chapters and submissions to this edited volume also attest to this fact. For example, eleven proposals on virtual worlds, from authors spanning five different countries, were submitted for consideration for publication in this book. Of those, two focused on formal learning outcomes, seven focused on informal learning outcomes, and two investigated the use and meaning of avatars. These proposals were submitted by individuals working both in industry and academia, and the submissions from academics came from fields as diverse as instructional design, teacher education, distance education, nursing, art education, and mathematics. This diversity is not limited to virtual worlds: a similar phenomenon was observed for proposals investigating wiki-related topics and Web 2.0 technologies.

Following from the thesis that emerging technologies transcend academic disciplines, it seems worthwhile to put forth an education-specific definition to guide our thinking, research, and practice. Establishing a common understanding of a widely used term represents the first step towards meaningful conversations and inquiry.

What Are Emerging Technologies?

First, a personal story. In the summer of 2008, I received an e-mail that announced the release of an open-access e-book while also noting that the editor was "editing a new series of which this book is the first. The series is entitled Issues in Distance Education and we welcome submissions or letters of interest from authors wishing to publish with an Open Access, peer-reviewed license." A few weeks later (and after contacting the series editor, press director, and lead editor), I was given permission to proceed with the edited volume that you are now reading. In the midst of completing my dissertation and moving to a different country for my first tenure-track appointment, I quickly found myself putting together a call for proposals (CFP) for an edited volume on the use of emerging technologies in distance education. In the next two months, I received more than sixty-five proposals. Emerging technologies in distance education seemed to be a "hot topic," and it seemed that we had managed to solicit chapter proposals at an opportune time. After acceptance/rejection decisions were made, I began writing the introduction to this book and decided to begin by quickly defining the term "emerging technologies." I scanned my personal bibliography. I typed the term in my favourite search engine. I searched the academic literature. To my amazement, a definition for the omnipresent term was elusive. I searched magazines, periodicals, and industry reports. I discovered a few descriptions, but no such thing as a formal, commonly accepted definition. I took it upon myself to define "emerging technologies" but quickly began doubting the absence of a definition. Could it be that a definition actually existed and I simply could not locate it?

I decided to ask my colleagues for assistance (Figure 1.1): I asked my Facebook friends; posted a working definition on my blog; e-mailed colleagues asking for the definition that they use, who in turn, posted the question on the online networks they frequent; and contacted all the authors whose papers appear in this volume. The answers I received were informative and shared some commonalities, but I could not find one single statement that uniformly explained the meaning of the term "emerging technologies." The term that was central to the book I was editing had never been defined, or, if it had been defined, neither I nor my expert colleagues were able to locate that definition.

Figure 1.1 Asking colleagues to offer their definition of the term *emerging technologies*

George is searching for a definition of "Emerging Technologies."
Anyone have one or know a source? 12:49pm · 12 Comments

anyone want to share their working definition of emerging technologies for teaching/learning? 2:26 PM Nov 26th from web

Definition of Emerging Technologies for Learning

I received an email recently asking for my definition of emerging technologies for learning. To enlarge the conversation, I asked the question on Twitter. The following are responses:

November 18, 2008

A definition of emerging technologies for education

Filed under: E-learning, Ideas, open, sharing, work — Tags: emergingtechnologies, emerging technologies, E-learning, new media, web 2.0, web2 — George Veletsianos @ 8:26 am Edit This

Surprisingly enough, the education, e-learning, educational technology, instructional design, and so on literatures do not include a definition of emerging technologies for education. Below is my attempt at defining the term. This definition

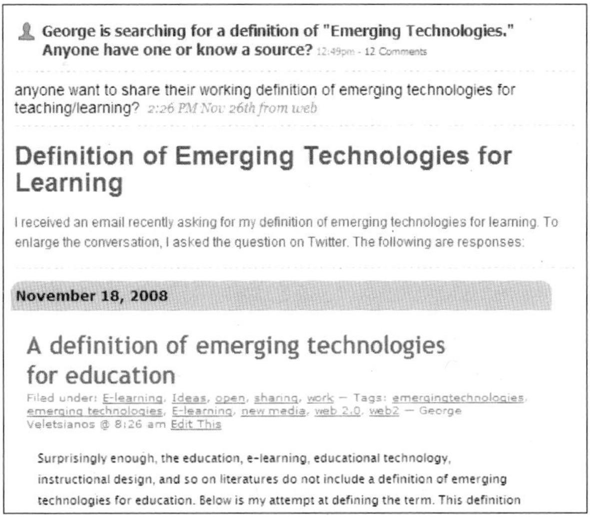

This experience provided the impetus for converting the book's short introduction into a chapter. How could a book on emerging technologies (in distance education or otherwise) exist without a shared understanding of what emerging technologies are?

At the same time, and since my initial search to discover a definition in the academic literature had proved futile, I focused on high-profile publications that specifically discussed emerging technologies for teaching and learning. The only explicit definition of emerging technologies I could locate in such publications came from a report commissioned for the Australian Capital Territory Department of Education and Training in which Miller, Green, and Putland (2005) state that

> A technology is still emerging if it is not yet a "must-have." For example, a few years ago email was an optional technology. In fact, it was limited in its effectiveness as a communication tool when only some people in an organization had regular access to it. Today, it is a must-have, must-use technology for most people in most organizations. In this sense a technology can be a standard expectation in the commercial or business world, while still being considered as "emerging" in the education sector. (p. 6)

Essentially, these authors note that any technology (defined as "infrastructures of various kinds, delivery devices, and classroom and teaching tools" on pp. 2 and 6) that is elective and not yet a requirement for educational organizations is considered to be an emerging technology. I find this definition to be an inadequate conceptualization of emerging technologies because it treats all technologies not currently used in educational institutions as emerging. While a number of technologies not currently in use in the education sector *may* be emerging, it is not necessarily true that *all* are emerging. Specifically, (a) organizations explore and adopt technologies even before they become "must-haves," (b) the notion of following others that popularize technologies as "must-haves" is problematic in that it implies that learning-focused organizations constantly follow on the footsteps of others, and (c) it disregards the potential of the technology for *educational purposes* — while some technologies may be "must-haves" for industries outside the educational realm, it does not necessarily mean that these same technologies are must-haves for educational providers. Finally, the notion of the specific situation one is facing (e.g., in terms of students, learner characteristics, institution, local realities, etc) in influencing what can and cannot be classified as an emerging technology is an important factor in considering whether technologies are emerging or otherwise — I explore this issue in the Implications section of the chapter.

Another set of publications investigated were the Horizon Reports (http://www.nmc.org/horizon). Since 2004, the New Media Consortium (NMC) and the EDUCAUSE Learning Initiative (ELI) have released their yearly Horizon Reports, which, in short, lay out adoption horizons for key emerging technologies likely to have an influence on education. The sections of these reports describing the concept of "emerging technologies" to date (2004–2008) are presented in Table 1.1. We can make three observations from these descriptions. First, the reports have consistently described emerging technologies as "likely to have a large impact ... on teaching, learning, or creative expression ... within three adoption horizons over the next one to five years" (2004–2008). Second, while the reports have focused on "higher education" for the period 2004–2007, the focus was broadened to "learning-focused organizations" in 2008. Third, the reports fluctuate as to the impact and expected magnitude of the impact that emerging technologies will/may have: emerging technologies are expected to become "very important" (2004), are expected to become "increasingly significant" (2005), will have "significant impact" (2006), will "impact" (2007), and will "enter mainstream use" (2008). While the descriptions of emerging technologies given in these reports are relatively stable across the project's lifespan, the differences in the descriptions from year to year provide additional insight into emerging technologies. From these descriptions and their differences, it can be inferred that emerging technologies are technologies that have not yet been widely adopted and that are expected to influence a variety

Table 1.1 "Emerging technologies" definitions as given in yearly Horizon Reports 2004–2008 (emphasis added)

Year	Definition
2008	"The annual Horizon Report describes the continuing work of the New Media Consortium (NMC)'s Horizon Project, a five-year qualitative research effort that seeks to identify and describe emerging technologies *likely to have a large impact on teaching, learning, or creative expression within learning-focused organizations...* The main sections of the report describe six emerging technologies or practices that *will likely enter mainstream use in learning-focused organizations within three adoption horizons over the next one to five years.* Also highlighted are a set of challenges and trends that will influence our choices in the same time frames." (2008, p. 3)
2007	"The annual Horizon Report describes the continuing work of the NMC's Horizon Project, a research-oriented effort that seeks to identify and describe emerging technologies *likely to have a large impact on teaching, learning, or creative expression within higher education...* The core of the report describes six areas of emerging technology that *will impact higher education within three adoption horizons over the next one to five years.*" (2007, p. 3)
2006	"The annual Horizon Report describes the continuing work of the NMC's Horizon Project, a research-oriented effort that seeks to identify and describe emerging technologies *likely to have a large impact on teaching, learning, or creative expression within higher education...* Each year, the report describes six areas of emerging technology that *will have significant impact in higher education within three adoption horizons over the next one to five years.*" (2006, p. 3)
2005	"The second edition of the NMC's annual Horizon Report describes the continued work of the NMC's Horizon Project, a research-oriented effort that seeks to identify and describe emerging technologies *likely to have a large impact on teaching, learning, or creative expression within higher education...* The report highlights six areas of emerging technology that the research suggests *will become increasingly significant to higher education within three adoption horizons over the next one to five years.*" (2005, p. 3)
2004	"This first edition of the NMC's annual Horizon Report details the recent findings of the NMC's Horizon Project, a research-oriented effort that seeks to identify and describe emerging technologies *likely to have a large impact on teaching, learning, or creative expression within higher education...* The 2004 Horizon Report ... highlights six technologies that the research suggests *will become very important to higher education within three adoption horizons over the next one to five years.*" (2004, p. 2)

Source: New Media Consortium (NMC) and the EDUCAUSE Learning Initiative (ELI). Reports retrieved 2 November 2008, from http://www.nmc.org/horizon

of educational organizations within a time span of one to five years. The differences between the descriptions of expected impact across 2004–2008 point to the uncertainty that exists with regards to (a) whether these technologies will actually have an impact, and (b) the magnitude and importance of the expected impact. These differences are important because as the next section describes, *uncertainty* is an important aspect of emerging technologies.

The Horizon Reports' definitions of the term *emerging technologies* seem to encompass the main ideas of what we traditionally consider to be "emerging technologies," but the fluctuations in terms of expected impact are problematic. Additionally, the reports focus on "teaching, learning, or creative expression" even though emerging technologies may potentially alter organizational structures, and influence leadership and scholarship.

The third report studied is entitled "Emerging technologies for learning." This is a publication of the British Educational Communications and Technology Agency (BECTA) that has also sought to understand the implications of emerging technologies. The introduction to the three editions of the "Emerging technologies for learning" reports (retrieved 28 October 2009, from http://bit.ly/147D9C) states that the publication

> aims to help readers consider how emerging technologies may impact on education in the medium term. The publications are not intended to be a comprehensive review of educational technologies, but offer some highlights across the broad spectrum of developments and trends. It should open readers up to some of the possibilities that are developing and the potential for technology to transform our ways of working, learning and interacting over the next three to five years.

As is the case with the Horizon Reports, BECTA emphasizes the *possibility* of a near-future impact. Broader than the Horizon Reports, BECTA also sees emerging technologies influencing the way we work and interact.

A Comprehensive Definition of Emerging Technologies

To define the term *emerging technologies for education*, I explored how researchers and practitioners perceive these technologies, what their functions, characteristics, and affordances are perceived to be, and what is known and not known about them. Further, I attempted to define emerging technologies in terms of their properties and not in terms of the actual technologies that are categorized as emerging (e.g., Web 2.0 technologies are often considered to be emerging technologies, and while the two terms are often used synonymously, I treated them as being distinct; more accurately, I attempted to define emerging technologies without focusing on features of Web 2.0). This process led to an initial definition that was then questioned through the ideas and definitions offered by others (presented as a mind map in Figure 1.2). These contributions acted as a peer-review system for my initial thoughts and research while lending further credibility to the definition I offer below.

Figure 1.2 A mind map of ideas offered by other researchers and practitioners when the question was asked on Twitter, Facebook, in e-mails, and on various blogs.

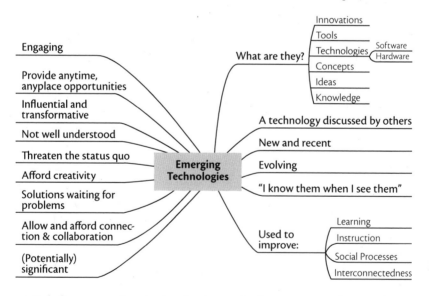

I define emerging technologies as tools, concepts, innovations, and advancements utilized in diverse educational settings (including distance, face-to-face, and hybrid forms of education) to serve varied education-related purposes (e.g., instructional, social, and organizational goals). Emerging technologies (ET) can be defined and understood in the context of the following five characteristics:

1. Emerging technologies may or may not be new technologies.

It is important to note that the words *emerging* and *new* are usually treated as synonymous, but they may not necessarily be so. While a definition of *new* might be perilous and contentious, ET may represent *newer* developments (e.g., utilizing the motion-sensing capabilities of a video-game controller to practice surgical techniques) as well as older ones (e.g., employing open-source learning management systems at higher-education institutions). Even though it may be true that most emerging technologies are newer technologies, the mere fact that they are new does not necessarily categorize them as emerging. If we were to treat new technologies as emerging technologies, the following two questions would arise:

When do technologies cease to be *new*?
When technologies cease to be *new*, do they also cease to be emerging?

For example, synthetic (or virtual) worlds were described as an emerging technology in the mid-1990s (Dede, 1996), with research on Multi-User Dungeons dating back to the 1980s (Mazar & Nolan, 2008). Yet, virtual worlds are still described as emerging technologies (e.g., see chapter 15 and de Freitas, 2008). *Newness,* by itself, is a problematic indicator of what emerging technologies are; *older* technologies can also be emerging, and the reasons for this will become clearer after we examine the characteristics that follow.

2. Emerging technologies are evolving organisms that exist in a state of "coming into being."

The word *evolving* describes a dynamic state of change where technologies and practices are in a continuous state of refinement and development. To illustrate this, consider the chalkboard. We no longer have discussions about how to use the chalkboard, and even though dry-erase boards may be less dusty and easier to use, the way the board is used is generally established; to a large extent, the community has agreed and settled on the use of the chalk/dry-erase board. On the other hand, Twitter, the currently popular social networking and micro-blogging platform, represents an illustrative example of an ET that is "coming into being." For example, Twitter's early success and popularity caused frequent outages, and such issues were most noticeable during popular technology events (e.g., during the 2008 MacWorld keynote address). Early attempts to satisfy sudden surges in demand included using more servers and implementing on/off switches on various Twitter features (e.g., during the 2008 Worldwide Developers Conference), while later efforts included re-designing the application's architecture and withdrawing services (e.g., free SMS and instant-messaging support). Existing in a state of evolution, Twitter has engineers who continuously develop and refine the service, while maintaining its core purpose. On the whole, Twitter exists in an evolutionary state where new features change the way the technology is used, and new users engage in practices that may depart from what was originally anticipated. Nevertheless, and it is important to note this, while Twitter may be an emerging technology, various practices and activities on the Twitter platform may already be established. An example of this is the ReTweet (RT) activity (boyd, Golder, & Lotan, 2010).

3. Emerging technologies go through hype cycles.

Today's emerging technology might be tomorrow's fad, and today's simple idea might be tomorrow's key to boosting learner engagement or university outreach. While it is easy to fall into the trap of believing that today's innovations will restructure and revolutionize the way we learn and teach, it is important to maintain skepticism towards promises of sudden transformation. Even though technology has had a major impact on how distance education is delivered, managed, negotiated, and practiced, it is also important to recognize that due to organizational, cultural, and historical factors, education, as a field of study and practice, has been relatively resistant to change (see Cuban, 1993; Lortie, 1975).

Technologies and ideas go through cycles of euphoria, adoption, activity and use, maturity, impact, enthusiasm, and even infatuation. In the end, some of today's emerging technologies (and ideas) will become staples, while others will fade into the background. One way to describe the hype that surrounds emerging technologies and ideas for education is to observe the Hype Cycle model (Fenn & Raskino, 2008) developed by Gartner Inc. This model evaluates the relative maturity and impact of technologies and ideas, and follows five hype stages: Technology Trigger, Peak of Inflated Expectations, Trough of Disillusionment, Slope of Enlightenment, and Plateau of Productivity (Fenn & Raskino, 2008). Most specific to the topic of this book are the hype cycle models developed for higher education (Gartner, 2008b), e-learning (Gartner, 2006), and emerging technologies (Gartner, 2008a).

4. Emerging technologies satisfy two "not yet" criteria:

(a) Emerging technologies are not yet fully understood.

One factor distinguishing emerging technologies from other forms of technology is the fact that we are not yet able to understand the implications of these technologies. What do they offer to education, teaching, and learning? What do they mean for learners, instructors, and institutions? For example, what exactly does socialization via social networking sites mean for distance learners? What does it mean to have "access" to others via an "add as friend" button? Could social networking sites break down digital divides? Or are social networking sites another medium through which societal inequalities are perpetuated? What are the pedagogical affordances of social networking sites? How can we sustain learner engagement in online learning communities? Can location-aware devices enhance communal learning experiences? As a result of emerging technologies not being fully understood, a second issue arises:

(b) Emerging technologies are not yet fully researched or researched in a mature way.

Initial investigations of emerging technologies are often evangelical and describe superficial issues of the technology (e.g., benefits and drawbacks) without focusing on understanding the affordances of the technology and how those affordances can provide different (and hopefully better) ways to learn and teach at a distance. Additionally, due to the evolutionary nature of these technologies, the research that characterizes them falls under the case study and formative evaluation approaches (Dede, 1996), which is not necessarily a drawback, but it does point to our initial attempts to understand the technology and its possibilities. Because emerging technologies are not yet fully researched, initial deployments of emerging technology applications replicate familiar processes, leading critics to argue that technologies are new iterations of the media debate (e.g., Choi & Clark, 2006; Clark, 1994; Kozma, 1994; Tracey & Hasting, 2005), without an understanding of the negotiated and symbiotic relationship between pedagogy and technology. Yet such criticisms are not entirely misplaced: newer technologies are often used in old and familiar ways. For example, linear PowerPoint slides replace slideshow projectors; blogs — despite the opportunities they offer for collaboration — replace personal reflection diaries; and pedagogical agent lectures replace non-agent lectures.

5. Emerging technologies are potentially disruptive but their potential is mostly unfulfilled.

Individuals and organizations may recognize that a potential exists within a technology, but such potential has not yet been realized. The reasons may be found in the features of emerging technologies already discussed. For instance, education is relatively resistant to change and mature research has not yet been conducted on the numerous emerging technologies used. Lack of research impedes dissemination and diffusion. Additionally, the potential to transform practices, processes, and institutions, is both enthusiastically welcomed and ardently opposed. A well-known example with regards to the open education movement concerns open-access journals. Supporters claim that free and open access has the potential to transform the ways research and knowledge are disseminated and evaluated. While this advancement has the potential to disrupt scholarship, to date — for a number of reasons — the majority of research is still published in fee- and subscription-based journals and periodicals, even though institutions are slowly moving towards open-access repositories (see http://dash.harvard.edu/ and https://www.escholar.manchester.ac.uk/).

To summarize, emerging technologies are tools, concepts, innovations, and advancements utilized in diverse educational settings to serve varied education-related purposes. Emerging technologies

(1) may or may not be new technologies,
(2) can be described as evolving organisms that exist in a state of "coming into being,"
(3) experience hype cycles,
(4) satisfy the "not yet" criteria of
    (a) not yet being fully understood, and
    (b) not yet being fully research or research in a mature way, and
(5) are potentially disruptive, but their potential is mostly unfulfilled.

Implications

The proposed definition provides a glimpse into the complexities that arise when emerging technologies are utilized in educational contexts. Although educational practitioners and researchers may consider emerging technologies powerful instruments in our quest to enhance teaching, learning, student engagement, and educational systems worldwide, we are still learners, still learning what is possible to achieve with these technologies. The absence of a large empirical or practitioner knowledge base to guide our work is evident. Rather than viewing this issue as a drawback, however, I would like to see it as an opportunity to explore how we can enhance educational practice. We should remain open to the idea that existing ways of teaching and designing learning environments may not serve the twenty-first century purposes of education. Note that I am not arguing that students are "wired" differently due to technological exposure or that we should abandon sound pedagogical principles. On the contrary, what we do know about learning, teaching, and education from such diverse fields as psychology, instructional design, sociology, and neuroscience, is important in our quest to understand and utilize emerging technologies. At the same time, technology is changing the way we live and act in the world (e.g., browsing physical books is a completely different experience when we can evaluate the quality of a book as a result of viewing online reviews received though augmented-reality software). Employing emerging technologies to further educational goals may necessitate the development of different theories, pedagogies, and approaches to teaching, learning, assessment, and organization. If we employ emerging technologies in our work, we should also be prepared to experiment with different lenses through which to view the world and with different ways to explore such ideas and practices as knowledge, scholarship, collaboration, and even education. While doing so, we should also remain cognizant of the fact that resistance and failures are possible, and, if documented in the literature, helpful. A few advances on this front have already been undertaken, and they include *connectivism* as an example of a learning theory to capitalize on networked knowledge (Siemens, 2005) and *social network knowledge construction* as an example of a pedagogical approach that enables instructors to integrate social network technologies into learning environments (Dawley, 2009).

The proposed definition of emerging technologies also implies that technologies cannot be seen as being "emerging" out of context. More specifically, technologies may be emerging in one area, while already being established in another area. For example, geographic information systems may already be established tools in the real estate and agriculture industries, but they are still considered to be emerging in the teaching of K–12 geography (Doering & Veletsianos, 2007). Perhaps more importantly, a technology may be established *and* emerging within the same field, at the same time. For example, interactive whiteboards are already established and pervasive in the United Kingdom's primary and secondary school sectors (BECTA, 2006; Hall & Higgins, 2005; Kennewell & Higgins, 2007). The scene is different at higher-education institutions: interactive whiteboards, while mostly available in teacher-training departments, are still in a state of emergence with instructors struggling to devise ways to use them (Brown, 2003). Within the field of education, therefore, interactive whiteboards are, at the same time, both emerging and established. Finally, in an e-mail message concerning the proposed definition, Alec Couros (author of chapter 6) pointed out that the contextual nature of emerging technologies might also hold true for differences across nations, regions, and even organizations. Examples include countries bypassing landline infrastructure and "leapfrogging" to mobile technologies when others, such as Canada, cannot support mobile technologies due to heavy regulation and geography; citywide wireless Internet for some cities (e.g., Minneapolis, MN) while not for others (e.g., Brainerd, MN); and the use of technology to support problem-based teaching techniques in one classroom in a K–12 school as compared to using technology for drill-and-practice exercises in a different classroom within the same school.

The link between emerging technologies as "evolving organisms that exist in a state of 'coming into being'" and the sociological theory of emergence (see Clayton, 2006) was highlighted by Hagit Meishar-Tal (author of chapter 11) in a private e-mail, and is also discussed in Whitworth and Benson's chapter (chapter 10). Emergent theory posits that events and phenomena do not happen in a formal or predetermined way, but rather, occur spontaneously and unexpectedly in dynamic environments that both influence activities and are influenced by the activity (Cole & Engestrom, 1993; Moje & Lewis, 2007). The implications of emergent theory for emerging technologies in education are twofold: on the one hand, technologies developed for purposes other than education find their way into educational institutions and processes (e.g., wikis), while on the other, once such technologies are integrated into educational practice, they both mould and are moulded by micro-educational practices, such as teaching and learning activities and communities of practice (chapter 10).

Finally, it is important to highlight that, in addition to categorizing various *tools* as emerging technologies, the definition of the term allows it to also describe ideas, theories, and approaches. This is only natural. In the same way that the word *technology* arises from the Greek word *techne* (τέχνη), meaning "craft" or "art," emerging technologies encompass both the tools *and* the ideas that are emerging and emergent. Examples of approaches that can be described as emerging technologies are: adventure learning, an approach to the design of authentic, experiential, and collaborative adventure-based learning environments (chapter 5); the utilization of the personal learning network within an open teaching framework (chapter 6); and the use of artistic pedagogical technologies (chapter 7).

Concluding Thoughts

In 2007 the Association of Educational Communications and Technology returned to the use of the term "educational technology" to define a field that, over the years, has been referred to by numerous names, including "instructional design," "instructional systems," and "instructional systems technology" (Reiser, 2006). In response to the name change, Lowenthal and Wilson (2009) argued that definitions and labels are critically vital because they establish a common ground upon which we can have conversations. An agreed-upon definition can enable colleagues to discuss ideas and research upon a shared understanding, enabling the field to move forward. Without an agreed-upon definition, the very foundations of our work are precarious.

The definition of emerging technologies for education provided in this chapter lays the foundations upon which to position our work. In addition to highlighting important issues for future research and practice, this definition also provides a starting point from which our work (and the rest of the chapters in this volume) can be conceptualized, extended, and evaluated.

REFERENCES

BECTA. (2006). *The BECTA Review 2006: Evidence on the Progress of ICT in Education*. British Educational Communications and Technology Agency, Coventry. Retrieved November 14, 2008, from http://publications.becta.org.uk/download.cfm?resID=25948

boyd, d., Golder, S., & Lotan, G. (2010). Tweet tweet retweet: Conversational aspects of retweeting on Twitter. *Proceedings of HICSS-43*. Kauai, HI: IEEE Computer Society. 5–8 January 2010. Retrieved 29 October 2009, from http://www.danah.org/TweetTweetRetweet.pdf

Brown, S. (2003). Interactive whiteboards in education. *TechLearn for Joint Information Systems Committee*. Retrieved 13 November 2008, from http://www.jisc.ac.uk/uploaded_documents/Interactivewhiteboards.pdf

Bull, G., Knezek, G., Roblyer, M.D., Schrum, L., & Thompson, A. (2005). A proactive approach to a research agenda for educational technology. *Journal of Research on Technology in Education, 37*(3), 217–220.

Choi, S., & Clark, R.E. (2006). Cognitive and affective benefits of an animated pedagogical agent for learning English as a second language. *Journal of Educational Computing Research, 34*(4), 441–466.

Clark, R.E. (1994). Media will never influence learning. *Educational Technology Research and Development, 42*(2), 21–29.

Clayton, P. (2006). Conceptual foundations of emergence theory. In P. Clayton & P. Davies (Eds.), *The Re-Emergecne of Emergence: The Emergentist Hypothesis from Science to Religion* (pp. 1–31). Oxford: Oxford University Press.

Cole, M., & Engestrom, Y. (1993). A cultural-historical approach to distributed cognition. In G. Salomon (Ed.), *Distributed Cognitions: Psychological and educational considerations* (pp 1–46). New York, NY: Cambridge University Press.

Cuban, L. (1993). *How Teachers Taught: Constancy and Change in American Classrooms, 1880–1990*. (2nd ed.). New York, NY: Teachers College Press.

Cuban, L. (2001). *Oversold and Underused: Reforming Schools through Technology 1980–2000*. Cambridge, MA: Harvard University Press.

Dawley, L. (2009.) Social network knowledge construction: Emerging virtual world pedagogy. *On the Horizon, 17*(2), 109–121.

de Freitas, S. (2008). Serious virtual worlds: A scoping study. *Joint Information Systems Committee*. JISC report. Retrieved 17 November 2008, from http://www.jisc.ac.uk/publications/publications/seriousvirtualworldsreport.aspx

Dede, C. (1996). Emerging technologies and distributed learning. *American Journal of Distance Education, 10*(2), 4–36.

Doering, A., & Veletsianos, G. (2007). An investigation of the use of real-time, authentic geospatial data in the K–12 classroom. *Journal of Geography, 106*(6), 217–225.

Fenn, J., & Raskino, M. (2008). *Mastering the Hype Cycle: How to Choose the Right Innovation at the Right Time*. Cambridge, MA: Harvard Business School Press.

Gartner Inc. (2006). *Hype Cycle for Higher E-Learning, 2006*. Retrieved 12 November 2008, from http://www.gartner.com/DisplayDocument?doc_cd=141123

Gartner Inc. (2008a). *Hype Cycle for Emerging Technologies, 2008*. Retrieved 12 November 2008, from http://www.gartner.com/DisplayDocument?doc_cd=159496

Gartner Inc. (2008b). *Hype Cycle for Higher Education, 2008*. Retrieved 12 November 2008, from http://www.gartner.com/DisplayDocument?doc_cd=158592

Hall, I., & Higgins, S. (2005). Primary school students' perception of interactive whiteboards, *Journal of Computer Assisted Learning, 21*(2), 102–117.

Kennewell, S., & Higgins, S. (2007). Introduction. Special issue, *Learning, Media and Technology, 32*(3), 207–212. Retrieved from http://www.informaworld. com/smpp/title%7Econtent=t713606301%7Edb=all%7Etab=issueslist%7 Ebranches=32-v32

Kozma, R. (1994). Will media influence learning? Reframing the debate. *Educational Technology, Research and Development, 42*(2), 7–19.

Lortie, D. (1975). *Schoolteacher: A Sociological Study.* Chicago: University of Chicago Press.

Lowenthal, P., & Wilson, B.G. (2009). Labels DO Matter! A Critique of AECT's Redefinition of the Field. *TechTrends, 54*(1), 38–46.

Mazar, R., & Nolan, J. (2008). Hacking say and reviving ELIZA: Lessons from virtual environments. *Innovate, 5*(2). Retrieved 3 December 2008, from http://www. innovateonline.info/index.php?view=article&id=547

Miller, J., Green, I., & Putland, G. (2005). *Emerging Technologies: A Framework for Thinking.* Australian Capital Territory Department of Education and Training. Retrieved 13 November 2008, from http://www.det.act.gov.au/__data/ assets/pdf_file/0019/17830/emergingtechnologies.pdf

Moje, E.B. & Lewis, C. (2007). Examining opportunities to learn literacy: The role of critical sociocultural research. In C. Lewis, P. Enciso, & E.B. Moje (Eds.), *Reframing Sociocultural Research on Literacy: Identity, Agency, and Power.* Mahwah, NJ: Lawrence Erlbaum Associates.

Reiser, R.A. (2006). What field did you say you were in? Defining and naming our field. In Reiser, R.A., & Dempsey, J.V. (Eds), *Trends and Issues in Instructional Design and Technology* (pp. 2–9). Upper Saddle River, NJ: Merrill/Prentice Hall.

Siemens, G. (2005). Connectivism: A learning theory for the digital age. *International Journal of Instructional Technology and Distance Learning, 2*(1). Retrieved 21 September 2008, from http://www.itdl.org/Journal/Jan_05/ article01.htm

Siemens, G. (2008). Complexity, Chaos, and Emergence. Retrieved 11 November 2008, from https://docs.google.com/View?docid=anw8wkk6fjc_15cfmrctf8.

Tracey, M.W., & Hasting, N.B. (2005). Does media affect learning: Where are we now? *TechTrends, 49*(2), 28–30.

2 THEORIES FOR LEARNING WITH EMERGING TECHNOLOGIES

> *Terry Anderson*

Abstract

This chapter is designed to give an overview of some of the traditional and new learning and instructional theories that guide the effective development and deployment of emerging technologies in education. Theories force us to look deeply at big-picture issues and grapple with the reasons why our technology use is likely to enhance teaching and learning. This chapter provides an overview of various visions for the use of educational technology and learning theories associated with those visions, and concludes with a brief look at three modern, Net-centric theories of learning.

> Creating a new theory is not like destroying an old barn and erecting a skyscraper in its place. It is rather like climbing a mountain, gaining new and wider views, discovering unexpected connections between our starting points and its rich environment. But the point from which we started out still exists and can be seen, although it appears smaller and forms a tiny part of our broad view gained by the mastery of the obstacles on our adventurous way up. (Albert Einstein, in Einstein & Infield, 1938, 158–9)

Introduction

In this chapter I outline some of the most relevant established and emerging learning, pedagogical, and educational theories that both inspire and guide our interest in exploiting emerging technologies

23

for distance education applications. While educational theory is often construed by graduate students as a necessary evil of little practical use, required by professors and thesis committees, I have written elsewhere about the value of theory in education development and design (Anderson, 2004b). Summed up by Kurt Lewin's (1952) famous quote, "there is nothing so practical as a good theory" (p. 169), I begin this chapter with a short personal anecdote.

During the summer of 2003, I began to see a flood of new Web-based information and communications technologies that could be used to create learning activities in formal education. At that time, I became obsessed with the notion that there must be some sort of rational law that would help educators and instructional designers decide when to use which particular technology. Moreover, the mere fact that a technology is popular for personal or business use provides little evidence that it will be useful in educational contexts. In addition, I was worried (and still am) that the adoption of any new technology, in traditional contexts, is hard work, often disruptive, and will likely have unanticipated consequences. Thus, I was searching for theoretical constructs to guide interventions.

I was drawn to thinking about the technologies in the context of Moore's (1989) description of educational communications as being made up of student-student, student-content, and student-teacher interactions. We had already written (Anderson & Garrison, 1998) about the other three possible interactions — teacher-content, teacher-teacher, and content-content — but continued to focus on the ones most relevant to a learning centric view, those that involved students. I created the diagram shown in Figure 2.1 and then had an insight: perhaps these three student interactions were more or less equivalent. Creating very high-quality levels of any one type of interaction would be sufficient to create a high-quality learning experience. If this was the case, the other two interactions could then be reduced or even eliminated, with little or no impact on learning outcomes or learner attitudes. If true, this "learning equivalency theory" could be used to rationalize expenditures in one area, yet allow for time and money savings in the other two. I further speculated that "high levels of more than one of these three modes will likely provide a more satisfying educational experience, though these experiences may not be as cost- or time-effective as less interactive learning sequences" (Anderson, 2003).

Figure 2.1 Learning interactions

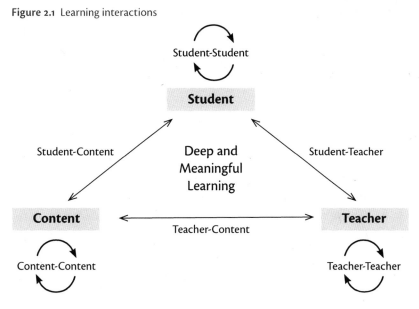

The problem with this "theory" rests on Popper's 1968 claim that a good theory is one that can never be proved true, but should be capable of being proved false. I had little idea how to disprove this theory and thus thought its contribution to the field might at best be as an interesting hypothesis and as a rubric for course designers. I was thus very pleasantly surprised to read that Bob Bernard (Bernard et al., 2009) and his colleagues at Concordia University, had thought deeper than I, and had established a set of protocols that allowed them to conduct a meta-analysis of distance education studies designed to validate my contentions. As usual, the number of control group studies in distance education is limited and thus so are the results. However, Bernard et al. (2009) concluded that "when the actual categories of strength were investigated through ANOVA, we found strong support for Anderson's hypothesis about achievement and less support for his hypothesis concerning attitudes."

Thus, my "equivalency theory" gained some empirical support, and from e-mails I have received from distance educators in a variety of countries, I know the theory has helped both researchers to research and practitioners to design and deliver cost-efficient and learning-effective interventions.

The remainder of this chapter reviews some of the older and newer theories that I find of most interest and value in my own thinking and practice, and I hope this overview helps the reader to understand and act effectively in the emerging world of online education that we are creating.

Historical Theories of Educational Technology

Good theories stand the test of time and continue to be of use because they help us understand and act appropriately. These theories are useful today because emerging technologies are often applied to the same challenges and problems that inspired educators and researchers working with older technologies, technologies that, while now established, were once emerging (chapter 1). As aptly stated by Larreamendy-Joerns and Leinhardt (2006), "the visionary promises and concerns that many current educators claim as novel actually have a past, one whose themes signal both continuities and ruptures."

In a fascinating review of educational technology research and its application to online learning, Larreamendy-Joerns and Leinhardt (2006, p. 568) define three views or visions that propel educational technology use and development. These are: the presentational view, the performance-tutoring view, and the epistemic-engagement view.

The *presentational view* focuses on theory and practice that make our discourse and especially our visualizations more clearly accessible to learners. Theories of multimedia use focus on the cognitive effects of selecting and transmitting relevant images and words, organizing these transmissions effectively, and insuring that the messages delivered through multiple channels do not interfere with each other or with the cognitive processing of the learners (Mayer, 2001). Much of this work benefits from studies of brain activity, and our increasing understanding of the complex ways in which we process "presentations" helps us to create these presentations in most effective ways.

The *performance tutoring* view derives its roots from the feedback, reinforcement, and theory of behavioural psychology. More recently, social constructivist theories have focused on the role of scaffolds provided by both human and non-human agents that assist more able or knowledgeable learners or teachers to prompt and support learners in acquiring their own competence (Vygotsky & Lauria, 1981).

Constructivism

The *epistemic engagement* view of learning has been the most recent educational vision driving educational technology. Engagement is most closely associated with constructivist learning theories. Currently the most popular approach to learning that is guiding both researchers and educational practitioners is a set of theories collectively known as *constructivism*. Constructivism has long philosophical and pedagogical roots and has been associated with the works of John Dewey, George Mead, and Jean Piaget. Like many popular theories, it has been defined and characterized by many — often with little consistency among authors. However, all forms of constructivism share an understanding that individuals construct knowledge that is dependent upon their individual and collective understandings, backgrounds, and proclivities. Debate arises, however, over the degree to which individuals hold common understandings and if these understandings are rooted in any single form of externally defined and objective reality (Kanuka & Anderson, 1999). As much as constructivism is touted as driving the current educational discussion, it should be noted that it is a philosophy of learning and not one of teaching. Despite this incongruence, many authors have extracted tenants of constructivist learning and from them developed principles or guidelines for the design of learning contexts and activities. Among these are: that active engagement by the learners is critically important, and that multiple perspectives and sustained dialogue lead to effective learning. Constructivists also stress the contextual nature of learning and argue that learning happens most effectively when the task and context are authentic and hold meaning for the learners.

Constructivist learning activities often focus on problems and require active inquiry techniques. These problems often work best when they are ill structured, open ended, and messy, forcing learners to go beyond formulaic solutions and to develop their capacity to develop effective problem-solving behaviours across multiple contexts.

Complexity Theory

Complexity theory — or more recently, the "science of complexity" — arose from the study of living systems, and has been attracting esoteric interest among a very wide variety of disciplines for the past two decades. Perhaps most familiar examples of complexity theory are those drawn from evolutionary study, where organisms (over time) adapt to and even modify complex environments, creating unusually stable, yet complex systems. In such systems one component of an ecosystem cannot be understood in isolation from the context or total environment in which it lives (for an example, see chapter 10). Complexity theory teaches us to look for the emergent behaviours that arise when autonomous, yet interdependent organisms interact with each other. In particular, theorists look for and attempt to predict "transformations or phase transitions that provide the markers for growth, change, or learning" (Horn, 2008). Complexity theorists are often at odds with positivist researchers and educators who attempt to eliminate or control all the variables that affect a learning transaction. Rather, complexity seeks to create learning activities that allow effective behaviour to emerge and evolve and ineffective ideas to be extinguished. Conversely, complexity theorists seek to understand features of the environment, and especially the social or structural norms or organizations we create that resist either overt or covert attempts at self-organization. McElroy (2000) notes that "the point at which emergent behaviours inexplicably arise, lies somewhere between order and chaos" (p. 196). This sweet spot has been called the "edge of chaos," where systems "exhibit wild bursts of creativity and produce new and novel behaviours at the level of the whole system ... complex systems innovate by producing spontaneous, systemic bouts of novelty out of which new patterns of behavior emerge" (McElroy, 2000, p. 196).

Implications of complexity theory for learning and for education operate on at least two levels. At the level of the individual learner, complexity theory, like constructivist theory, supports the learner's acquisition of skills and power such that he or she can articulate and achieve personal learning goals (chapters 6 and 9). By noting the presence of agents and structures that both support and impede the emergence of effective adaptive behaviour, individual learners are better able to influence and indeed survive in often threatening and always complex learning environments.

At the level of organization of either formal or informal learning, complexity theory highlights the social structures that we create to manage that learning. When these management functions begin to inhibit the emergence of positive adaptive behaviour or give birth and sustain behaviours that are not conducive to deep learning, we can expect negative results. Organizational structures should help us to surf at the "edge of chaos," not function to eliminate or constrain the creative potential of actors engaged at this juncture. Further, this understanding can guide us to create and manage these complex environments not with a goal of controlling or even completely understanding learning, but with a goal of creating systems in which learning emerges rapidly and profoundly. Complexity theory also encourages us to think of learning contexts (classrooms, online learning cohorts, etc.) as entities themselves. These entities can be healthy or sick; emerging, growing, or dying. By thinking at the systems level, reformers search for interventions that promote healthy adaptation and the emergence of cultures, tools, and languages that produce healthy human beings.

Finally, complexity theory helps us to understand and work with the inevitable unanticipated events that emerge when disruptive technologies are used in once stable systems (Christensen, 1997). Learning to surf this wave of equal opportunity and danger (and do it masterfully) becomes the goal of educational change agents.

The teaching and learning theories derived from all three of these pre-Net visions for technology-enhanced learning and related theories of learning still resonate with and add value to educators and researchers today. However, to paraphrase the syllogism that "the Net changes everything," I next turn to theories that have evolved since the development of the Web and deliberately exploit the affordances of this new context for teaching and learning.

Net-Aware Theories of Learning

The Net context creates an environment that is radically different from pre-Net contexts, yet of course carries evolutionary genes from previous cultures and technologies. In 2004 Denise Whitelock and I edited a special edition of the *Journal of Interactive Media in Education* that focused on the educational semantic web (Anderson & Whitelock, 2004). In the introduction to this issue, we provided an overview of three affordances of the Web, which I still believe define its value for teaching and learning. The first is the capacity for powerful yet very low-cost communications. This capacity forms the platform upon which "epistemic-engagement" visions of learning are instantiated. This communication may be engaged in synchronous, asynchronous, or near-synchronous (as in text messaging) modes. Communications may be expressed through text, voice, video, or even immersive interaction modes. These communication modes can also be combined in many creative ways. Communication artifacts can be stored, indexed, tagged, harvested, searched, and sorted. All of this capacity is available at low or (at least in parts of the world) affordable cost. Finally, net communications can be one to one, one to many, or many to many, with very little cost differentiation among the three modes. Thus, educators have moved away from a world in which communication was expensive, geographically restricted (often limited to those sharing the same classroom), and privileged (limited to those with production facilities). Moreover, net communications provide access to and empower those with hearing, movement, or visual impairments. These communications affordances obviously can be used in a multitude of ways in formal education and teaching (see chapters 4, 12, and 14). The recent emergence of social software sites affords learners the opportunity to seek and share questions, understandings, and resources, thus creating learner-organized tutoring and support opportunities (chapter 6). Perhaps most importantly, this communications capacity creates opportunities for many forms of collaborative informal and lifelong learning (Koper & Tattersall, 2004).

The second affordance we discussed in 2004 is that the Net creates a context that moves us from information and content scarcity to abundance. From early-learning object repositories to wide-scale distribution and production of Open Educational Resources from many networked sites, the Net provides learning content with many different display and presentation attributes. This content exists in many formats, and often uses multimedia to enhance its presentational value. Most exciting is the capacity for learners and teachers to add user-created content and to edit and enhance the work of others using produsage production modes (Bruns, 2008).

The third affordance we identified in 2004 has been less apparent, but still holds great promise for teaching and learning. This is the affordance of active and autonomous agents that can be set loose on the Net to gather, aggregate, synthesize, and filter the Net for content and communications that is relevant to individual and groups of learners and teachers. The educational semantic web still remains "just around the corner," and there have been serious methodological (Doctorow, 2001) and epistemological (Kalfoglou, Schorlemmer, & Walton, 2004) challenges to its emergence. However, there is an increasing number of applications that utilize autonomous agents (Anderson, 2004a; Sloep et al., 2004) to induce and support learning. The most visible of these applications are the search-engine algorithms that we all use to find and retrieve Net-based content, products, and services. By noting which sites are selected most often and which have the most established traffic and links, search engines calculate short lists of options to select from among the often millions of matches that are found — and as often as not the "correct" site appears among those on the first result screen. Through agents actively monitoring the Net, the links, and the collective actions of users, algorithms produce an intelligent guess as to the searcher's desired result — as well as a few targeted advertisements! Furthermore, agents monitoring these searches extract additional information that is used by marketers and social researchers to further understand our collective ideas, choices, and interests (Tancer, 2008), as well as by researchers and educators who want to further understand learner behavior in Net-based learning environments (chapter 12). Net-based agents will doubtlessly continue to add value to all three of the visions for educational technology, including presentation, performance-tutoring, and epistemic engagement.

Nevertheless, being in awe of stunning technical affordances does little to direct or help us to understand teaching and learning. For this, I end the chapter with overviews of three more recent learning theories that evolved in the technology-enhanced learning networked area.

The Pedagogy of Nearness

The first of these network-centric learning theories relates to the capacity of learning to flow seamlessly between online and face-to-face contexts. Mejias (2005) has argued for a new "pedagogy of nearness" in which online interaction, collaboration, and learning are neither valued nor devalued as compared to interactions with those near at hand. We have all noticed the ease with which some of us move between online and offline living. And more recently we see evidence that Net-infused learning does not entail desertion of our physical spaces, but rather serves to facilitate, document, and deepen place-based communication and relationships (Ellison, Steinfield, & Lampe, 2007). In a network-infused environment, we are "on our way to a more sustainable relationship with the world when we learn to inscribe our online experiences into larger systems of action meant to bring the epistemologically far near to us, and make the physically near relevant again" (Mejias, 2005b).

Mejias also argues that it is not only the nearness of face-to-face interaction that presents unique opportunities for teaching and learning but rather "we need to acknowledge the kind of insights (about ourselves, about our world) that can be gained through online experiences that cannot be gained through unmediated perception" (Mejias, 2005). Although digital networks have not accomplished the death of distance, our sense of both time and distance has been altered in as yet little-understood ways by the cost-effective reduction of barriers to both. Mejias' work points to the need for blended applications in which networks are used for teaching and learning when appropriate and which offer particular access, time shifting, or pedagogical advantage. Moreover, learners and teachers must develop literacies to act effectively in both online and offline contexts and be able to shift rapidly between them.

Heutagogy

The second net-centric theory reviewed was developed in Australia and was named by its authors, Hase and Kenyon (2000), as *heutagogy* after the Greek for *self*. Heutagogy has roots in the literature on self-directed learning and renounces the teacher dependency associated with both pedagogy and andragogy. Heutagogy extends control to the learner and sees the learner as the major development and control agent in his or her own learning (Hase & Kenyon, 2007). The self-determinism that defines heutagogical approaches to teaching and learning is seen as critical to life in the rapidly changing economy and cultures that characterize postmodern times. As Hase and Kenyon (2000) note, "heutagogy looks to the future in which knowing how to learn will be a fundamental skill given the pace of innovation and the changing structure of communities and workplaces." This future demands that education move beyond instructing and testing for learner competencies to allow and support learners in a journey to capacity rather than competency. Capacity includes being able to learn in new and unfamiliar contexts. Older models of competence test only the time-dependent achievement of the past. Instructional design for heutagogy learning veers away from prescriptive content to an exploration of problems that are relevant to students' lives (chapter 5). The teacher's role becomes one of facilitator and guide as students use a very wide set of resources (both online and traditional) to resolve problems and to gain personal understanding and capacity. Heutagogy's emphasis on self-direction and capacity of heutagogy focuses on the development of efficacy in utilizing the tools and information sources available on the Net.

Connectivism

The most recent network-centric theory was first developed by George Siemens, who coined the term *connectivism* and laid out a number of principles that define connected learning. Siemens argues that "we derive our competence from forming connections" and that our "capacity to know more is more critical than what is currently known" (Siemens, 2005). The metaphor of the network whose nodes consist of learning resources, machines that both store and generate information, and people, dominates connectivist learning. Learning occurs as individuals discover and build connections between nodes. Learning environments are thus created and used by individuals as they access information, process, filter, recommend, and apply that information with the aide of machines, peers, and experts in their learning networks. In the process of learning, they expand their own learning networks by creating useful and personalized knowledge and connecting it to the ideas and artifacts of others in their networks. Being able to see, navigate, and create connections between nodes becomes the goal of connectivist learning. Rather than learning facts and concepts, connectivism stresses learning how to create paths to knowledge when it is needed. Siemens also argues that knowledge and indeed learning itself can exist outside of a human being — in the databases, devices, tools, and communities within which a learner acts.

Additionally, connectivism sees the need for formal education to expand beyond classrooms and bounded learning management systems to embrace and to become involved with the informal. As Downes (2006) notes, "Learning … occurs in communities, where the practice of learning is the participation in the community. A learning activity is, in essence, a conversation undertaken between the learner and other members of the community. This conversation, in the Web 2.0 era, consists not only of words but of images, video, multimedia and more." Though often the topic of edublogger conversation, connectivism has yet to become widely accepted as the learning theory for the digital era as envisioned by Siemens and Downes. It has been criticized by Kerr (2007) for offering nothing new in learning theory that is not accounted for in earlier works (notably complexity theory and constructivism). Connectivism also seems to have trouble connecting to formal education. Kop and Hill (2008) note the lack of a substantive role for a teacher in connectivist theory and the requirements placed on the learner (in common with heutagogy) to be capable of and motivated to engage in very self-directed learning. Finally, Verhagen (2006) argues that connectivism is more a theory of curriculum (specifying what the goal of education should be and the way students should learn in that curriculum) than a theory of learning.

Obviously a goal of connectivist learning is to create new connections, and the classroom, or any bounded formal education system, is a relatively small context in which to build these connections. Connectivist theorists are interested in both allowing and stimulating learners to create new learning connections. In the process, learners increase the pools of expertise and resources they can draw from and thus increase their own social capital, as they become valued resources for others. Our own modest contribution to this need for expanded interactions within formal education has been to differentiate three important but substantively different contexts in which connectivist learning is employed (Dron & Anderson, 2007).

The first of these learning contexts is the familiar group. Groups (often referred to as "classes" in formal education) are secure places where students aggregate (in classroom or online) and proceed through a series of independent and collaborative learning activities. Groups tend to be closed environments, have strong leadership from a teacher or group owner, and (in formal education) be temporally bounded by an academic term. These synchronized activities result in learners supporting each other, and levels of trust can be built such that learners actively engage with and critique each other. In well-organized groups, considerable social, cognitive, and teaching presence is developed to create a community of inquiry (Garrison & Anderson, 2003). However, groups are also noted for the development of hidden curricula, constrictive and occasionally coercive acts, group think, and teacher dependency (Downes, 2006).

Thus, in our courses we are developing a second form of aggregation based on networks. Networked learning activities that expand connectivity beyond the Learning Management System (LMS) to allow both registered students, alumni, and the general public to engage in creating networked learning opportunities (Anderson, 2005). Network membership is much more fluid than that of groups, leadership is emergent rather than imposed, and networks easily expand and contract as learners find them of more or less use in solving particular problems. Networks are also less temporally bonded and may continue to exist long after formal study terminates.

The third aggregation we have referred to as "collectives." Learning in collectives involves aggregating and synthesizing the myriad activities that go on over the Net and applying knowledge gained by these aggregations to particular problems. For example, searching very large aggregations of resources, such as found in Google, YouTube, or del.icio.us, and filtering them for perceived value or use allows us to selectively mine the activities of thousands or tens of thousands of individuals. These filterings can be socially magnified through collaborative resource tagging services such as citeulike.org and diig.com. Collective activities carry with them the potential for contagion and privacy invasion, but at the same time they allow us to benefit from the traces, tracks, recommendations, and activities of others, thereby creating paths which allow us to connect more easily to valued human and digital learning resources. We have expanded this discussion elsewhere to explore learning activities best suited for the "Aggregations of the Many" (Dron & Anderson, 2007).

Conclusion

This brief overview is intended to illustrate that understanding learning and learning designs that use emerging technologies can be enhanced by looking through the lens of both older and emerging educational and learning theories. Much of our understanding of how and why learning happens and the best ways to design effective learning activities is enhanced when we work from theoretical models. The Net, with its new affordances, seems to speed up and accentuate many of the ideas found in pre-Net learning theories.

However, as much as theories add value, they also need to evolve to account for the affordances as well as for any disruptive (Christensen, Horn, & Johnson, 2008) and unanticipated consequences (Taleb, 2007) of their use in any context. We are witnessing the birth and refinement of learning theories that work under the assumption of the ubiquitous Net. Like Net culture itself, these theories borrow from and expand pre-Net ideas, while envisioning new ways that knowledge is created, shared, and adapted.

REFERENCES

Anderson, T. (2003). Getting the mix right: An updated and theoretical rationale for interaction. *International Review of Research in Open and Distance Learning, 4*(2). Retrieved December 2007, from http://www.irrodl.org/index.php/irrodl/article/view/149/708

Anderson, T. (2004a). The educational semantic web: A vision for the next phase of educational computing. *Educational Technology, 44*(5), 5–9.

Anderson, T. (2004b). Towards a theory of online learning. In T. Anderson & F. Elloumni (Eds.), *Theory and Practice of Online Learning* (pp. 271–294). Athabasca, AB: Athabasca University.

Anderson, T. (2005). Distance learning: Social software's killer app? Paper presented at the ODLAA Conference, Adelaide. Retrieved December 2005, from www.unisa.edu.au/odlaaconference/PPDF2s/13%20odlaa%20-%20Anderson.pdf

Anderson, T., & Garrison, D.R. (1998). Learning in a networked world: New roles and responsibilities. In C. Gibson (Ed.), *Distance Learners in Higher Education* (pp. 97–112). Madison, WI: Atwood Publishing.

Anderson, T., & Whitelock, D. (2004). The educational semantic web: Visioning and practicing the future of education. *Journal of Interactive Media in Education, 1*. Retrieved December 2007, from http://www-jime.open.ac.uk/2004/1

Bernard, R.M., Abrami, P.C., Borokhovski, E., Wade, C.A., Tamim, R.M., Surkes, M.A., & Bethel, E.C. (2009). A meta-analysis of three types of interaction treatments in distance education. *Review of Educational Research, 79*(3), 1243–1289.

Bruns, A. (2008). *Blogs, Wikipedia, Second Life, and Beyond: From Production to Produsage.* New York, NY: Lang.

Christensen, C. (1997). *The Innovator's Dilemma: When New Technologies Cause Great Firms to Fail.* Cambridge: Harvard University Press.

Christensen, C., Horn, M., & Johnson, C. (2008). *Disrupting Class: How Disruptive Innovation Will Change the Way the World Learns.* New York, NY: McGraw Hill.

Doctorow, C. (2001). Metacrap: Putting the torch to seven straw-men of the meta-utopia. Retrieved 10 June 2003, from http://www.well.com/~doctorow/metacrap.htm#0

Downes, S. (2006). Learning networks and connective knowledge. Posted on IT Forum (paper 92). Retrieved November, 2008, from http://it.coe.uga.edu/itforum/paper92/paper92.html

Dron, J., & Anderson, T. (2007). Collectives, networks and groups in social software for e-learning. Paper presented at the World Conference on E-Learning in Corporate, Government, Healthcare, and Higher Education, Quebec. Retrieved February 2008, from www.editlib.org/index.cfm/files/paper_26726.pdf

Einstein A. and Infield, L. (1938). *The Evolution of Physics*. New York: Simon and Schuster.

Ellison, N., Steinfield, C., & Lampe, C. (2007). The benefits of Facebook "friends": Social capital and college students' use of online social network sites. *Journal of Computer-Mediated Communication, 12*(4), 1143–1168. Retrieved December 2007, from http://www.blackwell-synergy.com/doi/full/10.1111/j.1083-6101.2007.00367.x

Garrison, D.R., & Anderson, T. (2003). *E-Learning in the 21st Century*. London: Routledge.

Hase, S., & Kenyon, C. (2000). From Andragogy to Heutagogy. *UltiBase*. Retrieved 28 December 2005, from ultibase.rmit.edu.au/Articles/dec00/hase2.htm

Hase, S., & Kenyon, C. (2007). Heutagogy: A child of complexity theory. *Complicity: An International Journal of Complexity and Education, 4*(1), 111–118. Retrieved March 2008, from www.complexityandeducation.ualberta.ca/COMPLICITY4/documents/Complicity_41k_HaseKenyon.pdf

Horn, J. (2008). Human research and complexity theory. *Educational Philosophy and Theory, 40*(1).

Kalfoglou, Y., Alani, H.; Schorlemmer, M., & Walton, C. (2004). On the emergent semantic web and overlooked issues. In S. McIlraith, D. Plexousakis, & F. van Harmelen (Eds.), *The Semantic Web –ISWC 2004* (pp. 576–590). Berlin: Springer.

Kanuka, H., & Anderson, T. (1999). Using constuctivism in technology-mediated learning: Constructing order out of the chaos in the literature. *Radical Pedagogy, 2*(1). Retrieved 5 September 2003, from http://radicalpedagogy.icaap.org/content/issue1_2/02kanuka1_2.html

Kerr, B. (2007). A Challenge to Connectivism. Transcript of Keynote Speech. Paper presented at the Online Connectivism Conference. Retrieved November, 2008, from http://ltc.umanitoba.ca/wiki/index.php?title=Kerr_Presentation

Kop, R., & Hill, A. (2008). Connectivism: Learning theory of the future or vestige of the past? *The International Review of Research in Open and Distance Learning, 9*(3). Retrieved November, 2008, from http://www.irrodl.org/index.php/irrodl/article/view/523/1103

Koper, R., & Tattersall, C. (2004). New directions for lifelong learning using network technologies. *British Journal of Educational Technology, 35*(6), 689–700.

Larreamendy-Joerns, J., & Leinhardt, G. (2006). Going the distance with online education. *Review of Educational Research, 76*(4), 567–605.

Lewin, K. (1952). *Field Theory in Social Science: Selected Theoretical Papers*. London: Tavistock.

Mayer, R. (2001). *Multimedia Learning*. Cambridge: Cambridge University Press.

McElroy, M. (2000). Integrating complexity theory, knowledge management and organizational learning. *Journal of Knowledge Management*, 4(3), 195–203. Retrieved 6 June 2006, from http://www.emeraldinsight.com/Insight/ViewContentServlet?Filename=Published/EmeraldFullTextArticle/Pdf/2300040301.pdf

Mejias, U. (2005a). Movable Distance: Technology, Nearness and Farness. ulises mejias' blog entry. Retrieved November, 2008, from http://blog.ulisesmejias.com/2005/01/20/movable-distance-technology-nearness-and-farness/

Mejias, U. (2005b). Reapproaching nearness: Online communication and its place in praxis. *First Monday, 10*(3). Retrieved November, 2005, from http://firstmonday.org/issues/issue10_3/mejias/index.html

Moore, M. (1989). Three types of interaction. *American Journal of Distance Education, 3*(2), 1–6.

Popper, K.R. (1968). *The Logic of Scientific Discovery*. New York: Harper & Row.

Siemens, G. (2005). A learning theory for the digital age. *Instructional Technology and Distance Education, 2*(1), 3–10. Retrieved October 2005, from http://www.elearnspace.org/Articles/connectivism.htm

Sloep, P., van Rosmalen, P., Brouns, F., van Bruggen, J., de Croock, M., Kester, L., & de Vries, F. (2004). Agent support for online learning. *Open Universiteit Nederland Educational Technology Expertise Centre*. Retrieved October 2004, from http://dspace.learningnetworks.org/retrieve/498/Agent_support_for_online_learning.pdf

Taleb, N. (2007). *The Black Swan: The Impact of the Highly Improbable*. New York, NY: Random House.

Tancer, B. (2008). *Click: Unexpected Insights for Business and Life*. New York, NY: Hyperion.

Verhagen, P. (2006). Connectivism: A new learning theory? *SurfSpace*. Retrieved November, 2008, from http://www.surfspace.nl/nl/Redactieomgeving/Publicaties/Documents/Connectivism%20a%20new%20theory.pdf

Vygotsky, L., & Lauria, A. (1981). The genesis of higher mental functions. In J.V. Wertsch (Ed.), *The Concept of Activity in Soviet Psychology* (pp. 144–188). Armonk, NY: Sharpe.

3 IMAGINING MULTI-ROLES IN WEB 2.0 DISTANCE EDUCATION

> *Elizabeth Wellburn & B.J. Eib*

Abstract

This chapter focuses on emerging technologies that we see as having the most profound impact on learning: online social communication/ Web 2.0 tools and environments. In the online social networking environment, individuals can switch seamlessly through varying roles of expert, amateur, audience, author, learner, and educator. Web 2.0 has redefined how information is created and shared, and is enabling the transformation of distance education. In this chapter we examine the world for our learners outside of their formal learning environments. In doing so, we question whether informal learning has changed things so profoundly that traditional approaches are becoming irrelevant, and we conclude by suggesting that educators should embrace a multiplicity of roles and, with our students and the general public, recognize and participate in dynamically and collaboratively constructed formal and informal personalized learning environments.

Part 1: Are we experts and amateurs, audience and authors, learners and educators — all at the same time? Perhaps Web 2.0 and our "role(s)" in distance education are causing us to reinvent ourselves.

Imagine the expert and the amateur

In the not-too-distant past, if we needed to learn something, we would almost certainly interact with an expert, either directly with a teacher, indirectly through a text document, or, less likely but also indirectly, through other forms of media. In any of those scenarios, the source of information was filtered before it reached us (e.g., our teachers had to have received a set of credentials, the newspaper or book would have been edited by someone with recognized expertise, etc.). Shirky (2008b) refers to this idea as the "filter then publish" model. If we eventually acquired enough information and received the appropriate degrees, we were then deemed as recognized experts ourselves, ready to be sought out by others.

Today, with many of us immersed as contributors and consumers of collaborative sources of information such as blogs, wikis, social networks, virtual worlds, and citizen journalism websites, the traditional notion of expertise is being questioned. Is it possible to ever acquire "enough" information? Which sources are to be trusted? The information supply from the Web 2.0 world has not, by definition, been vetted in any conventional sense of the word. Shirky calls this idea the "publish then filter" model (2008b).

Is the concept of expertise changing, or vanishing entirely? People without formal qualifications can contribute to the online information environment as easily as those who are recognized as experts; although it is almost a cliché to state that there is an explosion of online content, it probably bears repeating that with so much additional digital information being added daily. During a random four-day period in October 2009, Wikipedia added more than 4,000 articles, 6,500 pages, and a million edits (Wikipedia:About, n.d.); between May 2008 and May 2009, Facebook experienced a 97 percent increase in unique visitors and Twitter a 2,681 percent increase (Singer, 2009); and YouTube reports that every minute, twenty hours of video is uploaded to its site (YouTube, 2009), there is no guarantee that, when searching the Web, we will find information that has authoritative weight to it. Is this a problem for us as educators, or for education in general? If so, when is it a problem and when does it become a problem? What, if anything, can be done to address it?

Another feature of the not-too-distant past is that it was often difficult to acquire any degree of proficiency in areas outside of one's own field. We may have had the desire, but it was unlikely that the information was available to us. Hobbies were possible, but in-depth niche learning was only for the individual who had enough time and/or money to fully pursue an area of interest. Additionally, geographic, occupational, and socio-economic boundaries meant that a person was probably isolated from any community that could support his or her growth. Today, to use a phrase coined by Leadbeater (2005), a "passionate amateur" can easily engage with hobbies, interests, and academic and leisure pursuits in a way that is far beyond "dabbling," because (a) information is widely and cheaply accessible, and (b) the participatory nature of Web 2.0 means that a two-way information flow is available to all. Both amateurs and experts, and all those in between, can access information, collaborate, and network online with others who share similar interests/passions. Learning can be reciprocal, with experts learning from and building upon the ideas generated by non-experts. Examples are plentiful, from the recent story (Celizic, 2008) of a parent putting his child's medical records online to connect with researchers who might be able to work with him to help solve the puzzle of brain injury, to stories of citizen journalism exposing events that would have been otherwise hidden[1] to the point where law enforcers, politicians, and others can never assume that anything is "off the record" (Slocum, 2008). Amateurs are contributing in ways that were impossible a few years ago. Shirky uses Linux as an early open-source case study,[2] demonstrating the potential for enormous success through the "global talent pool." If participation is cheap, even for amateurs, then it's easy to experiment with a multitude of ideas. A small but dedicated group of people can easily find each other and cooperate on projects of common interest

Imagine the audience and the authors

Prior to Web 2.0, there was usually a clear distinction between an audience and a recognized author. The author was the rare individual who had enough information or talent to make it worthwhile financially to create an expensive publication; the audience was the rest of us who received that publication (or film, play, etc.). In contrast, Web

2.0 has also been referred to as the "read-write web" (O'Hear, 2007) where very cheap publication can ensue: the participatory capabilities of the most recent Internet tools such as wikis, blogs, etc., allow content to be contributed and viewed by anyone who has Web access. This means that small bits of information, generated by huge numbers of individuals, can be easily published to form vast information sources (e.g., Wikipedia). Shirky (2008b) poses the vision of a world where large numbers of people contribute massive amounts of knowledge to online collaborative projects (e.g., Wikimedia projects), even when their contribution takes up only small portions of their time, drawn from what he calls the cognitive surplus (for instance, time that may have previously been spent watching television commercials[3]). To an extent, large amounts of information are already abundant and easily and freely accessible. If we are not able to find the information we are searching for, we can request it (e.g., in a blog or micro-blogging platform) and it is possible that it will be generated for us (examples of such requests are presented in chapters 1 and 6). We can share our interpretations, we can comment, question, and critique information in a public sphere, and generate further conversations. Wikipedia is a clear example of how the author and the audience are one and the same, since everyone who reads Wikipedia articles is also provided with the ability to edit and write them, as well as make comments and engage in discussion with other participants.

Certain features of our widely used technologies are enabling us to have continuous access to information. Our cell/mobile phone is an Internet browser, and our computer is a telephone; we can send pictures and video clips instantaneously with the prospect of being viewed by millions, and we are easily able to listen to more voices than we've ever heard before. At our fingertips, at all times, the potential exists to be audience and author. It is therefore easy to become enthused if we know that it is simple to contribute, and that our small contributions can potentially be valuable. What questions could be solved with such vast intellectual capital directed towards them?

Imagine the learner and the educator

Like expert/amateur and audience/author, the roles of learner and educator are increasingly becoming intermingled in the Web 2.0 environment. Teachers may have always felt the pressure of keeping up-to-date in their field, but it is a profound change that both the learner and the "teacher" have identical access to the same vast set of resources. Even more of a dilemma is the possibility that the learner may have a potential advantage by being more familiar with the social networking aspects of information sharing (e.g., posing questions to online forums), and using skills acquired through gaming or sites such as Facebook. Downes (2008) discusses how such technologies have led to a more informal type of learning "based on a student's individual needs, rather than as predefined in a formal class, and based on a student's schedule, rather than that set by the institution." He goes on to describe how this informal learning involves "no boundaries; people drift into and out of the conversation as their knowledge and interests change." In our personal experiences with educators, we have noted that after their formal education has ended, they often tend to learn in a similar, "just in time" fashion, based on what is needed and drawing upon a range of relevant, but not predefined sources.

Oblinger and Oblinger (2005) refer to students who have grown up with technology as the "Net Generation." [4] Whether Net Gen learners are truly different from previous generations has since been debated. [5] (For instance, see the 2009 OECD report at http://www. nml-conference.be/wp-content/uploads/2009/09/NML-in-Higher-Education.pdf.) However, whether we apply the Net Gen label or not, an overall increase in access to technology cannot be disputed. A much-circulated Michael Wesch YouTube video (2008), "A Vision of Students Today," shows the viewpoint of learners (in their own words) within the traditional four-wall classroom and explores how the structured environment does not connect with their desire for informal learning and how the concept of categorized information does not fit with their ways of freely accessing what they need to know. These learners explicitly state that they hate school but love learning, they access social networking sites in class, they often don't read textbooks or assigned readings, they find school has a lack of relevance to life, and they don't see how multiple-choice questions will help them solve complex societal problems or allow them to succeed in a job that doesn't even exist yet; in the words of Perelman (1993), "school plods where human imagination naturally leaps."

As distance educators we are both part of, and separate from, the traditional education environment.[6] However, like almost all of today's educators, we have arrived here through a system that embraced neither the notions of informal learning nor of the expert, amateur, audience, and author in the relationships described above. As Garrison and Anderson stated in 2003, "unfortunately, the transmission model that still dominates education has changed little" (p. 1), and Robinson (2009) and Liston, Whitcomb, and Borko (2009), among others, note that there is still a reliance on this model wherever standardized testing is emphasized.[7] We have, however, likely used some technology, and perhaps even created online resources through a learning management system (LMS). Are we confident that we are on the right path, or are we apprehensive?

Do we fear that some learners have expectations that do not match what we want to provide? How does our curriculum match student expectations, if our learners expect informal learning? Should we even attempt to meet student expectations? Do we think we are the experts and they are the students who have come here to learn and that we should be the ones deciding what it is that they need to know? Or does informal learning prepare students for the future better than formal learning does? Will breadth and immediacy replace depth and analysis? What are the implications for the learning experiences we wish to create? Is the concept of the structured LMS still valid, or has it become obsolete? (See chapters 6, 9, and 10.) Should we be developing learning environments at all? Does our structured approach simply represent outdated views of how learning should take place, which should be abandoned in favour of a democracy of information composed of many small pieces from a range of sources (rather than a few large pieces from a small number of authoritative sources)? If and when an unstructured, student-led approach to teaching is preferable, what is the role of the educator? Are there situations where the structured approach is preferable and if so, how do we identify them?

A new responsibility seems to be upon us: to ensure that our learners have the opportunity to develop skills and literacies that are appropriate for deep learning from (or in spite of) the published but unfiltered information they are currently encountering. How do we fulfill this responsibility? How do we design in a way that anticipates what our learners will encounter in their futures? How do we ensure, as Siemens (2005) asked in his introduction of the concept of connectivism, that our learners develop the core skill of being able "to see connections between fields, ideas, and concepts?"

Part 2: What is the plan? How do we reinvent ourselves? How much will be different?

The participatory Web has elicited polar opposite views with respect to education and learning.

Some critics:

> Andrew Keen (2007), whose *The Cult of the* Amateur: *How Today's Internet is Killing Our Culture* expresses his concern regarding the watering-down of the concept of expertise and the flood of misinformation.

> Nicholas Carr (2008), whose article "Is Google making us stupid?" argues that hyperlinked reading on the Web has led us to be unable to focus on lengthier ideas, such as those in books.

> Christine Rosen (2008), whose article "The myth of multitasking" describes her concerns about multitasking with references to neurological changes and loss of productivity.

Some enthusiasts:

> Clay Shirky (2008b), whose book *Here Comes Everybody* can be summarized as: how Web 2.0 finally allows us to contribute collectively for the improvement of all by better using our cognitive surplus.

> John Seely Brown and Richard Adler (2008), in whose article "Minds on fire: Open education, the long tail, and learning 2.0" the

supply-push model (a factory model that relates older teaching strategies as building up inventory in students' heads) is compared to the demand-pull model (which is learning 2.0 or learning on demand) or "passion-based" learning. These authors argue that understanding is socially constructed and that meaning is created "by what one person produces and others build on — a remix."

> Stephen Downes (2008), whose article "The future of online learning: Ten years on" shares his continuing vision of self-directing and self-motivated learners and education as an act of liberty, possible only because collaborative technologies now allow fully participatory worldwide learning communities.

How do we find guidance in this diversity of opinion, especially when the evidence surrounding the use of emerging technologies in education is limited (chapter 1), and, as is often the case in educational research, mixed? The authors of this chapter work in an environment of distance educators and designers who are inspired by the enthusiasts who see the collective knowledge pool as an advancement in human culture. But we also recognize that others are concerned about over-optimism and have viewpoints similar to those expressed by Keen and Carr with regards to the indiscriminate blending of author and audience and the potential confusion that may result. If we accept that there is some validity in both points of view, we should feel compelled to explore the ways in which educators can work with (rather than fight against) what learners bring to educational pursuits so that their formal learning experiences afford them with an improved ability to evaluate and contribute at a more meaningful level. Perhaps now the challenges are:

(1) how to find ways to embed or scaffold critical thinking through the use of technology in general, and Web 2.0 tools and emerging technologies in particular;

(2) how to best assist our learners to be effective participants in the participatory society and to add value to the world they are living in; and

(3) how to advance distance education and enhance practice.

Shirky (2008a) states that "the physics of participation is more like weather than gravity. All the forces combine." It's a quote that evokes images of chaos: powerful but complicated patterns with unpredictable global consequences, compared to what he seems to see as our previous "what goes up must come down" way of looking at the world. Applied to distance education, if even a small part of what Shirky is imagining about this change is true (for instance, if we see a truth in his observation that the online world has flipped us from "filter then publish" to "publish then filter"), then it seems clear that teaching and learning must also be in transition. Wesch (2008) goes as far as to say that his every assumption about information and learning was shattered because of 2.0. *Shattered* is a very strong word, implying destruction, yet it is clear that in his own work, he has found a way to pick up the pieces. As distance educators, can *we* see any shattered pieces and find delight that some of our constraints have been lifted so we can refocus, rebuild, and reinvent?

The Wesch (2008) lecture at the University of Manitoba, "A portal to media literacy," is an excellent place to start thinking about reinventing ourselves within the distance education context. Wesch speaks in a lecture hall and bases his discussion of traditional education on that physical environment. He describes the hall as a place designed to fit a model of learning that incorporates the following beliefs:

(1) To learn is to acquire information.
(2) Information is scarce (so a place must be created where an expert can convey information to a large group).
(3) The authority of the expert must be followed (that is why the expert is at the front of the room with everyone else facing him/her).
(4) Authorized information is beyond discussion (so the chairs are in fixed positions and learners don't turn to talk to each other).

Wesch then describes varying interactions with learners, including using discussions and wikis to solicit their opinions, and his findings that learners no longer believe in the above assumptions. He concludes that there is a serious crisis of significance. His answer is to encourage learners to work on collaborative projects, and to use media tools for the making of meaningful connections with personal relevance. Assessment should be based on a view of whether and how learners have made those personally relevant connections rather than on the recitation of factual information.

Looking at the problems Wesch identifies, viewed in the context of the potential of Web 2.0 Distance Education, gives us a hopeful perspective regarding much of what is disengaging to lecture-hall learners in the information era. Distance Education has already taken important steps towards providing rich and relevant learning environments through the new tools.

Table 3.1

Wesch describing the lecture hall:	Contrasting concepts from our online Distance Education experience:
To learn is to acquire information.	To learn is to achieve learning outcomes, which may include gathering, analyzing, and evaluating ideas and building upon them to create new ideas and products.
Information is scarce (so a place must be created where an expert can convey information to a large group).	The online instructor is a facilitator, but does not lecture. Readings, videos, cases, etc., are provided, but there is an expectation that learners will find and share additional sources of information.
The authority of the expert must be followed (that is why the expert is at the front of the room with everyone else facing him/her).	There is no physical space so the attention is focused on whoever is providing the most relevant information at any given point in time.
Authorized information is beyond discussion (so the chairs are in fixed positions and learners don't turn to talk to each other).	There are no chairs, but we have provided online discussion forum areas, wikis, chat rooms, etc., for learners to share ideas.

How are we transitioning to a new model of distance education? Largely through the use of online media with appropriate pedagogies and ways of thinking about education and learning. To Wesch, media literacy is an important key to effective education in the Web 2.0 learning environment. He states, "There are no natives." Given that the online media environment is largely new to both educators and learners, Wesch suggests that we must not assume students are media literate. As an example, he mentions that a large proportion of his students did not know that Wikipedia was editable and many had never edited a wiki of any sort. And since emerging media and new tools are appearing nearly every day, media literacy strategies are more important than specific details about specific platforms. Other authors agree: Alexander (2008) argues that higher education must rethink the definition of literacy, "if we want our students to engage the world as critical, informed people, then we need to reshape our plans as that world changes" (p. 200). Wesch (2009) speaks of critical analysis and metacognition and of ways in which he engages students to create notes collaboratively, all related to his view that it's important to prepare students to create content in and for a world that is both "download and upload." Based on what his students are telling him, he believes that discussion (in our view, *critical* discussion or *true dialogue*; see chapter 14), rather than information transmission, is a key factor for engagement, and states that "the focus is not on providing answers to be memorized, but on creating a learning environment more conducive to producing the types of questions that ask students to challenge their taken-for-granted assumptions and see their own underlying biases" (paragraph 28).

How does the key idea of critical discussion, of engaged learning through posing questions and discussing ideas, position distance education? The early history of distance education was often a story of isolation (Sherry, 1996). Many who lived in areas too remote for schools to be accessible, were too ill, or could not afford to attend regular classes could learn alone, with workbooks and assignments exchanged through postal mail. An occasional telephone conversation with an instructor might have been included, but solitary learning was a fundamental and central feature of the early "correspondence" model. It seemed that the correspondence model was accepted as satisfactory and generally seen as second-best when compared to face-to-face learning. For instance, Garrison wrote in 1990 that the quality of the educational process depended upon two-way communication and he asserted that without connectivity, distance learning "degenerates" into the correspondence course model of independent study. The earliest distance education technologies were unidirectional and asynchronous (e.g., radio and broadcast television) and did not incorporate interaction (see chapter 2). When technologies able to diminish isolation and provide interaction opportunities became available (e.g., Scardamalia & Bereiter, 1994), distance learners and educators felt excited.[8] Distance education may be well positioned to be at the forefront of learning via Web 2.0, simply because there is little nostalgia for the early ways of teaching, studying, and learning in isolation. Having few compelling reasons to hold on to old methods does not mean that we don't face substantial hurdles in learning to harness the new possibilities; rather, it means that we need to envision new solutions for the current and future challenges.

There are more ways than one to achieve a learning outcome

Those distance educators familiar with the newer LMS environments (e.g., Moodle, BlackBoard) have probably incorporated discussion forums and collaborative assignments into their courses and may believe that our environments are (a) better than correspondence courses, and (b) not as limiting as a lecture hall. Hopefully, many of us are looking for ways to capitalize on this, to exploit the potential of the technology even further. Does our curriculum allow for utilizing technology to engage learners? Curricula may not always explicitly utilize technology in such ways, but there are more ways than one to achieve a learning outcome. Are we engaging learners by ensuring their learning is personally relevant? If not, could blogging or building a wiki for a real audience assist with this? Or, would it seem relevant for teams of learners to contribute to their favourite topic on Wikipedia, after researching the history trail on that topic to see how it has evolved? Do we assess on the basis of meaningful connections? Or would it make sense to ask our learners to create an online portfolio that demonstrates how they connect ideas? Should we view our learners' preparatory work done in collaborative bookmarking sites as evidence of a process used for connecting ideas? By the end of their distance education experience, will learners internalize and exhibit an enhanced ability to contribute to what John Seely Brown (2008) would call an "open-source culture," or create more of what Putnam (2000) would refer to as "social capital"? Or should we ask learners to launch a campaign on a social networking site to solve a complex societal problem as proof of the concept that online engagement can make a difference in the real world?

What about the risks suggested by the critics?

Shirky (2007) counters Carr's argument (that we are not reading as deeply in the era of abundance) by stating that "every past technology I know of that has increased the number of producers and consumers of written material, from the alphabet and papyrus to the telegraph and the paperback, has been good for humanity." Although Shirky sees Web 2.0 and information abundance as providing an opportunity for us to create and solve problems using collective cognitive surplus, Keen worries that we will falter by having too much freedom and too much access to information not created by recognized experts. Without sarcasm, Shirky agrees that Keen poses a hard question that must be answered: "What are we going to do about the negative effects of freedom?" Andy Carvin (2008) makes a similar suggestion when asking educators to avoid the "wide-eyed cheerleader" point of view and recognize the challenges.

Part of the solution may come from the technology itself. In the near future, there may well be technologies that evolve to provide authority to certain information. For example, Internet founding father, Tim Berners-Lee (2008, interviewed by Ghosh) is working on a project to provide scientific websites with reliability ratings, something he sees as being crucial for particular types of content (e.g., medical information/advice). But in general, as Keohane (2008) notes about Wikipedia, and by association Web 2.0, user-generated content is largely self-correcting. This point is profoundly important to the discussion.

In many ways the critics are calling for critical thinking and a type of virtual "street smartness." What is required are ways to ensure that user self-correction is ongoing and that users keep track of where any particular piece of information might be in that self-correction process (e.g., the first iteration of a Wikipedia article may be suspect; after a thousand edits, it may well be a highly reliable source). Without that awareness, the perils are indeed real. With awareness, the potential, in the view of all but the harshest critics, could be truly amazing. Whatever the case, there is probably no turning back. While recognizing and respecting the concerns of the critics, we should move forward with a spirit of adventure, applying our imagination and inventiveness to authentic questions.

It has long been discussed in K–12 education that learning should be authentic (e.g., Brown, Collins, & Duguid, 1989). Instead of merely studying history, learners should become historians, emulating the research techniques used by experts. They should learn science by doing science, and so on. We believe that authentic learning is increasingly made possible by the participatory nature of the Web 2.0 environment. If, as critics such as Carr, Keen, and others suggest, the inability to filter is one of the greatest arguments for avoiding Web 2.0 sources of information (thus staying with the model of "experts only" as content providers), then authentic learning provides a strong counterargument. When a consumer knows what's involved in creation, and is, in fact, a creator able to use the same techniques that experts use, there is a much smaller possibility that he or she will ever be misled. Authentic learning requires critical thinking based on experience. A simple example relates to image editing software. If you've ever used a digital image editing tool to delete, add, and replace people's heads and bodies, you know how easy it is to create a pictorial illusion; thus, you may be much more skeptical when you see images on the Web. Critical thinking skills become even more important in a world where professionals can create illusions such as those used in the food and fashion industries (e.g., http://tinyurl.com/56uefl).

Downes (2008) notes that the focus of the personal learning environment (chapter 9) "is more on creation and communication than it is consumption and completion…. We have seen the emergence of a new model, where education is practiced in the community as a whole, by individuals studying personal curricula at their own pace, guided and assisted by community facilitators, online instructors and experts around the world." These views are echoed by Brown (2008), who states that "social learning concerns not only 'learning about' the subject matter but also 'learning to be' full participants in the field." If you have been part of a community on a social networking site, asked or answered a question in an online forum, or built something in a virtual world, you've engaged in authentic learning and interactions that may help you interpret and filter information from similar contributors. You have harnessed your own personal learning environment.

Conclusion

Years ago, we heard about young people in China using fax machines to get information out beyond national borders (May, 1999). In recent years we have seen numerous examples where the combination of purpose or need with the use of social media and new Internet-based collaborative tools have significantly impacted world events and ways that people perceive these events.

The historic election of Barack Obama, whose successful campaign was driven by a Web 2.0 engine (an innovative Internet fundraising system that eventually attracted more than three million donors [Lister, 2008]), is one example. The fact that social software generated an unprecedented broad base of small donations indicates a new democratization of how political messages are given and received, and we are convinced that it reflects something truly disruptive about the core of Web 2.0 that confirms our belief regarding its impact in education and beyond.

News of the 2009 Iranian elections and the subsequent protests permeated Twitter for weeks. Hashtags were created and circulated, Twitter avatars were given a green tint to show solidarity with and support for the protesters, and some people even changed their Twitter profiles, claiming to be in the Iranian timezone in hopes that would make it more difficult for Iranian authorities to use tweets to identify protest organizers.

Web 2.0 technologies serve to easily and democratically connect people who may have previously had little or no opportunity to connect with each other, and this can occur regardless of the level of expertise of the participants. Such connections, based on interest and passion, foster new roles for learning, teaching, knowledge creation and knowledge consumption, supporting the social learning described by Brown and Adler (2008), who observe that "we participate, therefore we are" (p. 18) and describe projects such as the Faulkes Telescope project, which allows distant learners to "access scarce high-level tools" and "collaborate with working scientists" (p. 24).

As distance educators we can take on multiple roles through Web 2.0. Our learners, and the general public, can also take on multiple roles. Perhaps Web 2.0 and whatever comes next will enable us to reinvent our learning environments so that they are dynamically constructed with our learners and can include the greater public to become engaging and collaborative places of ongoing formal and informal personalized learning. We have exciting and fulfilling times ahead if we can adjust and participate.

NOTES

1 Regarding television, Shirky notes that U.S. television viewers spend "100 million hours every weekend, just watching the ads." (http://www.shirky.com/herecomeseverybody/2008/04/looking-for-the-mouse.html). A similar calculation based on the time a Canadian audience of 100,000 (probably only a portion of the total) spends watching commercials during a single hockey game would yield twenty-five full-time employees working for an entire year.

2 Linux is open-source software that was developed based on suggestions solicited through an early bulletin-board style discussion forum, and in Shirky's words it has "single-handedly kept Microsoft from dominating the server market" (p. 238).

3 Regarding television, Shirky notes that in the U.S., it's [What is? Need a clear subject and a more precise verb. Should be "an audience of 100,000 spends" or something?] "100 million hours every weekend, just watch

ing the ads." (http://www.shirky.com/herecomeseverybody/2008/04/looking-for-the-mouse.html). A similar calculation based on the time a Canadian audience of 100,000 (probably only a portion of the total) spends watching commercials during a single hockey game would yield twenty-five full-time employees working for an entire year.

4   Dede (2008) also views the new learning styles as having some profound differences that need to be accommodated, while also noting that over-simplification is possible when considering these learners.

5   Mark Bullen maintains a blog called Net Gen Skeptic (http://www.net-genskeptic.com) where he references the recent OECD report, stating "there is not enough empirical evidence yet to support that students' use of technology and digital media is transforming the way in which they learn, their social values and lifestyles, and finally their expectations about teaching and learning in higher education."

   And in the blog, Bullen mentions that "the report does conclude that students in higher education are heavy users of digital media and that they favour the use of technology but that they value technology use in education for its ability to improve access, convenience and productivity, not to radically change teaching and learning."

6   The distinction between face-to-face and distance education is becoming narrower. Integration of technology in the traditional classroom has allowed traditional classes to move to a hybrid model while distance educators have increasingly incorporated synchronous sessions.

7   Sir Ken Robinson says that "all the schools I know that are great have something in common — they all have great teachers and they have a commitment to the personal development of each of the pupils in the school. And that's easily lost in a culture of standardizing." (TEDBlog. http://blog.ted.com/2009/08/ted_and_reddit_1.php)

   Liston et al. (2009) refer to teachers working in the context of the U.S. No Child Left Behind Act, stating "greater emphasis is placed on preparing teachers who can get students to pass states' high-stakes assessments. Teacher preparation time is limited, and credit hours sometimes drastically reduced. Time spent has to be justified carefully and usually with an eye to K–12 student test scores" (p. 108).

8   One of the authors of this article has a personal recollection of the excitement of attending a training session showcasing Scardamalia and Bereiter's (1994) CSILE project as a model for what was possible in K–12 distributed learning.

REFERENCES

Alexander, B., (2008) "Social Networking in Higher Education." In Richard N. Katz, ed., *The Tower and the Cloud: Higher Education in the Age of Cloud Computing*. Educause: 2008.

Brown, J.S. (2008). How to connect technology and passion in the service of learning. *The Chronicle of Higher Education*. Retrieved 15 November 2008, from http://chronicle.com/weekly/v55/i08/08a09901.htm

Brown, J.S., & Adler, R.P. (2008). Minds on fire: Open education, the long tail, and learning 2.0. Retrieved 15 November 2008, from the Educause website: http://connect.educause.edu/Library/EDUCAUSE+Review/Mindson FireOpenEducationt/45823

Brown, J.S., Collins, A., & Duguid, P. (1989). Situated cognition and the culture of learning. *Educational Researcher, 18*(1), 32–42.

Carr, N. (2008). Is Google making us stupid? *The Atlantic*. Retrieved 15 November 2008, from http://www.theatlantic.com/doc/200807/google

Carvin, A. (2008) Web 2.0 and education: Hot or not? Retrieved 15 November 2008, from the Learning Now website: http://www.pbs.org/teachers/learning.now/2008/01/web_20_and_education_hot_or_no.html

Celizic, M. (2008). Father of brain-injured child offers hope to others. Retrieved 15 November 2008, from The Today Show website: http://www.msnbc.msn.com/id/27717674/

Dede, C. (2008). Planning for neomillennial learning styles: Implications for investments in technology and faculty. Retrieved 15 November 2008, from the Educause website: http://www.educause.edu/PlanningforNeomillen nialLearningStyles%3AImplicationsforInvestmentsinTechnologyandFaculty /6069

Downes, S. (2008) The future of online learning: Ten years on. Message posted to the Half an Hour Blogspot website: http://halfanhour.blogspot.com/2008/ 11/future-of-online-learning-ten-years-on_16.html

Garrison, D.R. (1990). An analysis and evaluation of audio teleconferencing to facilitate education at a distance. *The American Journal of Distance Education, 4*(3), 16–23.

Garrison, D.R., & Anderson, T. (2003) *E-learning in the 21st Century: A Framework for Research and Practice*. London: RoutledgeFalmer.

Ghosh, P. (2008). Warning sounded on Web's future. Retrieved 20 November 2008, from the BBC News website: http://news.bbc.co.uk/2/hi/technology/7613201.stm

Keen, A. (2007). *The Cult of the* Amateur: *How Today's Internet is Killing Our Culture*. New York, NY: Doubleday.

Keohane, K. (2008). Unpopular opinion: Everyone's an expert on the Internet. Is that such a bad thing? Retrieved November 20, 2008, from the AllBuUsiness website: http://www.allbusiness.com/technology/software-services-applications-internet-social/10589879-1.html

Leadbeater, C. (2005). The rise of the amateur professional (Online video clip). The TED.com talks. Retrieved November 22, 2008, from http://www.ted.com/index.php/talks/charles_leadbeater_on_innovation.html

Lister, R. (2008). *Why Obama Won*. Retrieved November 5, 2008, from the BBC News website: http://news.bbc.co.uk/2/hi/americas/us_elections_2008/7704360.stm

Liston, D., Whitcomb, J., & Borko, H. (2009). The end of education in teacher education: Thoughts on reclaiming the role of social foundations in teacher education. *Journal of Teacher Education, 60*(2), 107–111.

May, G. (1999, 19 February). Spamming for Freedom. Retrieved 1 December 2009, from The Nixon Center website: http://www.nixoncenter.org/index.cfm?action=showpage&page=wp9

Oblinger D., & Oblinger, G. (2005). Educating the net generation. Retrieved 15 November 2008, from the Educause website: http://www.educause.edu/educatingthenetgen/5989

O'Hear, S. (2007). E-learning 2.0: How Web technologies are shaping education. Retrieved November 5, 2008, from the ReadWriteWeb website: http://www.readwriteweb.com/archives/e-learning_20.php

Perelman, L.J. (1993). *School's Out: A Radical New Formula for the Revitalization of America's Educational System*. New York, NY: Avon.

Putnam, R.D. (2000). *Bowling Alone: The Collapse and Revival of American Community*. New York, NY: Simon and Schuster.

Robinson, K. (2009). TED and Reddit asked Sir Ken Robinson anything—and he answered. Retrieved 18 October 2009 from http://blog.ted.com/2009/08/ted_and_reddit_1.php

Rosen, C. (2008). The myth of multitasking. *The New Atlantis*. Retrieved 5 November 2008, from http://www.thenewatlantis.com/publications/the-myth-of-multitasking

Scardamalia, M., & Bereiter, C. (1994). Computer support for knowledge-building communities. *Journal of the Learning Sciences, 3*(3), 265–283.

Sherry, L. (1996). Issues in distance learning. *International Journal of Educational Telecommunications, 1*(4), 337–365.

Shirky, C. (2007). What are we going to say about "Cult of the Amateur"? Retrieved 15 November 2008, from the Many2Many website: http://many.corante.com/archives/2007/05/24/what_are_we_going_to_say_about_cult_of_the_amateur.php

Shirky, C. (2008a). Gin, television, and social surplus. Here Comes Everybody. Retrieved 15 November 2008, from http://www.shirky.com/herecomeseverybody/2008/04/looking-for-the-mouse.html

Shirky, C. (2008b). *Here Comes Everybody: The Power of Organizing Without Organizations.* New York, NY: The Penguin Press.

Siemens, G. (2005). Connectivism: A learning theory for the digital age. Retrieved 15 November 2008, from the *International Journal of Instructional Technology and Distance Learning* website: http://www.itdl.org/Journal/Jan_05/article01.htm

Singer, A. (2009). Social media, Web 2.0 and Internet stats. Message posted to The Future Buzz blog: http://thefuturebuzz.com/2009/01/12/social-media-web-20-internet-numbers-stats/. Retrieved 18 October 2009.

Slocum, Z. (2008). Web 2.0 Summit videos: Huffington, Musk, Gore. Retrieved 15 November 2008, from The Huffington News website: CNET News. http://news.cnet.com/8301-17939_109-10092190-2.html

Wesch, M. (2008). A portal to media literacy [Online video clip]. Retrieved 22 November 2008, from YouTube: http://www.youtube.com/watch?v=J4yA pagnros&feature=user

Wesch, M. (2009). From knowledgable to knowledge-able: Learning in new media environments. Retrieved 21 January 2009, from the Academic Commons website: http://www.academiccommons.org/commons/essay/knowledgable-knowledge-able

Wikipedia:About. (n.d.). *Wikipedia.* Retrieved 18 October 2009 from http://en.wikipedia.org/wiki/Wikipedia:About

YouTube. (2009). Traffic and stats. Retrieved 19 October 2009 from YouTube Fact Sheet: http://www.youtube.com/t/fact_sheet

BEYOND DISTANCE AND

4 TIME CONSTRAINTS:

Applying Social Networking Tools and Web 2.0 Approaches in Distance Education

> *Mark J.W. Lee & Catherine McLoughlin*

Abstract

This chapter assesses the value of Web 2.0 and its constituent social software tools in enhancing learning opportunities for distance students and addressing the traditional problems of distance education by enabling a greater sense of presence, community building, and participation. With reference to a number of examples of current innovative practices of distance educators, the chapter also outlines the transformative value of these emerging digital technologies, and signals emergent forms of learning and teaching that make use of the affordances of the new tools. The chapter focuses on three considerations that the authors believe are needed to effectively capitalize on the new possibilities of Web 2.0: (1) the use of social networking tools to build social presence; (2) the reconceptualization of the design approaches used to create and implement e-learning activities in distance education contexts; and (3) the consideration of pedagogical strategies used to support distance learners.

Introduction

"Web 2.0" (O'Reilly, 2005) is commonly used to describe an apparent second generation or improved form of the World Wide Web that emphasizes collaboration and sharing of knowledge and content among users. Characteristic of Web 2.0 are the socially based tools and systems referred to collectively as *social software*, which includes but is not limited to Web logs (blogs), wikis, Really Simple Syndication (RSS) and podcasting feeds, peer-to-peer (P2P) media sharing applications, and social bookmarking utilities. These new tools make possible a new wave of online behavior, distributed collaboration, and social interaction, and are already having a transformative effect on society, triggering changes in how we communicate and learn.

The uptake of Web 2.0 and social software tools is gaining momentum in all sectors of the education industry. In particular, Web 2.0 is seen to hold tremendous potential for addressing the needs of distance students, enhancing their learning experiences through increased connectivity, customization, personalization, and rich opportunities for networking and collaboration. Several authors and researchers have adopted this perspective as they make a case for a new understanding of distance teaching and learning (see, for example, chapters 2, 5, and 6; Anderson, 2005, 2007; Dron, 2007; Shih, Li, & Yang, 2007). Emerging social networking technologies now offer richer and greater possibilities for people to connect, share ideas, and participate in global communities than were previously available. In combination with appropriate learning designs and pedagogical strategies, these technologies hold enormous promise for enhancing, enriching, and extending traditional paradigms of distance education.

The present chapter commences with a review of the longstanding issues and problems of distance education, before considering how emerging social software technologies might be used to mitigate these issues, enhancing learning opportunities for distance students and enabling a greater sense of presence, community building, and participation. Drawing on a number of recent international examples of innovative online learning practices, the chapter also highlights how digital technology and Web 2.0 social networks are serving as catalysts for pedagogical change. By fuelling a move towards new forms and conceptualizations of distance teaching and learning, the new tools promote the facilitation of socio-experiential and authentic learning experiences for distance learners that support their needs and are aligned with the demands and challenges of the knowledge era and networked society.

The Loneliness and Isolation of the Distance Education Student

Distance learners have been shown to have the highest risk of dropping out of their programs of study at tertiary education institutions (Peters, 1992), a phenomenon that can be largely attributed to the isolation experienced by these students (Delahoussaye & Zemke, 2001; Hipp, 1997; Lake, 1999; Okun, Benin, & Brandt-Williams 1996; Peters, 1992; Rogers, 1990). Students desire a sense that they are part of a larger university community, rather than simply being an enrollee or statistic in a course. For many on-campus students, their involvement in the campus community forms an integral part of their social lives and plays an important role in their personal and academic development.

The "distance" factor inherent in distance education has been identified as one of the major problems for students studying in this mode (Meacham & Evans, 1989; Suen & Parkes, 1996). This geographical isolation significantly detracts from the need for social interactions that are usually afforded by face-to-face situations. On top of the practical problems of contacting academic and administrative staff, obtaining study materials, and gaining immediate access to resources such as laboratory equipment and library books, distance learners endure the disadvantage of being unable to interact in person with other students, which can put a significant damper on their motivation and enthusiasm for study. As such, they are very often denied a sense of belonging to a scholarly community (Galusha, 1997).

Another related concern for the distance learner is the perceived lack of contact with and timely feedback from an instructor. To an even greater extent than other students, distance learners are likely to have insecurities about learning (Knapper, 1988), and need both a level of guidance as well as assurance that they are on the right track. Without this support, they may face difficulty in self-evaluating their progress and their understanding of the subject material. Time management can become a problem as they invest inordinate amounts of their study time in activities deemed unimportant or less important by the instructor, or in futile searches for answers to queries that could have been clarified or resolved in a matter of minutes by asking a simple verbal question. Such issues can lead to considerable frustration with the distance education experience, and result in feelings of inadequacy as well as a lack of self-confidence (Wood, 1995).

Over the last few decades, large numbers of mature distance students have been entering universities with little idea of the institution's culture and few avenues enabling them to acculturate (West & Hore, 1989). According to Lake (1999), these students include "recyclers" seeking to upgrade their vocational or industry qualifications; "deferrers," who failed to take up offers of university places upon graduation from high school; "returners," who discontinued their initial university studies, often as a result of perceived isolation; and "early school leavers," who typically have negative memories of their past educational experiences. In his seminal work on distance education, Keegan (1996) asserts that the separation of student and teacher removes a vital link of communication between the two parties, which must be restored by means of explicit steps to "re-integrate" the teacher–learner interaction, albeit somewhat artificially, through measures such as ongoing electronic or telephone communication. In the absence of these measures, distance students are less likely to undergo acculturation into institutional life and are more likely to drop out (Sheets, 1992).

Another enduring issue in the distance education literature is the criticality of factoring into account the significant proportion of students who enroll with little or no experience in studying in this mode. This problem is compounded by the fact that many of these students may have had little or no experience with tertiary study in general, or have had prolonged absences from study. Unless they quickly develop academic "survival skills," these students are at considerable risk of withdrawing or failing (Wood, 1995). Of particular importance is the design of distance study materials and learning activities (Meacham & Evans, 1989; Race, 2005; Simonson, Smaldino, Albright, & Zvacek, 2005), which must carefully consider the special needs of these students.

Galusha (1997) does an excellent job at painting a broad overall picture of the abovementioned and other issues by listing six major categories of problems from the distance student's perspective:

> balance between costs (monetary and time) and motivators;
> availability of feedback and teacher contact;
> access to student support and services;
> feelings of isolation and alienation;
> lack of experience (in tertiary study and/or studying at a distance); and
> lack of (technical) training.

Garrison (1997) emphasizes the importance of social presence, which he proposes is the extent to which remote communicators can project themselves to others using any given technology or medium. Much research has been devoted to the creation and maintenance of social presence in technology-mediated distance learning environments (Rourke, Anderson, Archer, & Garrison, 1999). Without this dimension of connectivity between learners, in addition to teacher presence and rapport, distance learners will often flounder, become increasingly frustrated, and may ultimately withdraw or fail. The concept of social presence was first identified by Short, Williams, and Christie (1976), who defined it as the perception that one is communicating with people rather than with inanimate objects, despite being separated by geographical distance. The tendency and preference for people to work together in groups is a central tenet of social presence theory, so the model is of great interest to distance educators. According to Short et al., when social presence levels are low, group members feel disconnected, social cohesion is lessened, and group dynamics are weaker. Conversely, when social presence is high, members tend to feel more connected and engaged, and are motivated to participate in group processes such as collaborative learning. Research also shows that both individuals and groups will be better placed to accept technology-mediated communication as a substitute for face-to-face communication if social presence is high.

E-learning environments are now widely used to provide services to distance learners, and are capable of affording interactions that are needed to ensure learner-centered instruction and create a sense of social presence; with emerging Web 2.0 and social software tools, the potential is greater than ever before. The remainder of the chapter considers how the new wave of Web 2.0 and social software tools may be used to design, enhance, and deliver distance education, alleviating the problems of loneliness and isolation experienced by many students studying at a distance, and promoting high levels of social presence in the online environment. The emphasis is on exemplifying models and approaches to distance learning that capitalize on the affordances of social computing tools to create personalized, socially engaging, and connected learning experiences for distance learners.

Designing Authentic and Relevant Learning Spaces and Experiences for Distance Students

The task of designing high-quality technology-supported learning experiences is a significant challenge for educators (Bennett, Agostinho, Lockyer, & Harper, 2009; Lockyer, Bennett, Agostinho, & Harper, 2008), as it entails application of instructional design principles and knowledge of how learners operate in the online environment in order to create the optimum conditions for learning. Herrington, Reeves, and Oliver (2006) describe the task of designing for online and distance learning as a particularly complex process that involves fostering synergies among "learner," "task," and "technology." However, the challenge of creating engaging and immersive distance learning environments runs counter to the widespread practice of incorporating traditional classroom pedagogical strategies into the Web-based delivery of courses, for example, through a learning management system (LMS). Most widely accepted models of online higher education appear to entail reductionist approaches whereby LMSs are used to design easily digested packets of information, usually assessed by discrete, stand-alone tests and academic assignments. In contrast, Herrington et al. describe a model for the development of authentic tasks that can assist in designing environments of increased, rather than reduced, complexity. It provides a robust framework for the design of online, distance, and hybrid courses, based on the work of theorists and researchers in situated learning and authentic learning (Herrington et al., 2006; Oliver, Herrington, & Reeves, 2006).

The authentic learning framework describes characteristics of task design where it is the students who make the important decisions about why, how, and in what order they investigate a problem and learn the required skills. Distance educators and curriculum designers world-wide have adopted the model, as it considers the particular need of distance learners for self-directed and self-regulated learning. Many of these learners also have substantial professional and life experience, and therefore bring to distance learning encounters a wealth of prior knowledge, abilities, enterprise, and resources. Authentic learning offers a means of addressing the needs and expectations of these students for learning that is meaningful, relevant, and applicable to their personal and professional lives.

While Herrington et al. do not consider the issue of social presence specifically, they emphasize that tasks must be set in real-world contexts and require students to collaborate meaningfully, engage in peer evaluation, and connect with mentors and buddies in order to engender social and cognitive support. Although preceding the Web 2.0 revolution, their work shows significant promise and could be adapted to learning design in Web 2.0 contexts. However, Conole et al. (2008) assert that there is an inherent tension between the rhetoric surrounding the potential of Web 2.0 technologies and actual practice. The principles inherent in Web 2.0 are about the user, i.e., active participation, citizen journalism, the power of the network, and user-generated content, and yet few of these are currently being applied by educational designers or incorporated into innovative designs for learning.

As Web 2.0 offers an array of tools and affordances for sharing photos, media, and bookmarks, people develop shared interests in these objects and have conversations around them. Engeström (2005a, 2005b) defines this trend as "object-centered sociality," and the concept helps us to understand how Web 2.0 tools, the activities they facilitate, and the artifacts produced using these tools might be used in distance education settings to generate social networking and learning conversations. Bouman et al. (2007) have developed a design framework based on sociality. The principles they propose are intended to guide the design process and to ensure that learning environments enable the development of identities. They also suggest using metaphors and structures that resemble real life so that participants (learners) can identify with the activities associated with them. In the categories that apply to individuals, Bouman et al. indicate that building trust and relationships and enabling conversational interaction, networks, and feedback processes are fundamental to the success of the online learning experience. Through the use of social objects and spaces, people maintain convivial relationships and share ideas. Many of today's most popular and successful websites are built around the creation and sharing of social objects, for example, Flickr (photos), YouTube (videos), and Delicious (bookmarks). Weller (2008) explains that these social objects are valuable insofar as they stimulate and support conversational interaction, thereby creating a sense of immediacy among learners. In distance learning settings, there is a need to create tasks, content, and learning episodes that enable conversation and dialogue and the building of social rapport. These principles resonate with and extend the earlier work of Rourke and Anderson (2002) on the necessity of social, cognitive, and teacher presence in online communities needed to support distance learners.

The power and diversity of Web 2.0 tools is proving attractive for learners, who want to engage in immersive, participatory, socially involved, multi-modal experiences (Jenkins, 2007). Yet designers need to be cautious, as Moore (2007) indicates that "the overall effect of the new technology will be negative and counterproductive, if interest in the technology draws attention further from the need for reform in the way we design our courses and the need for better training and monitoring of instructors ..." (p. 182). A recent approach to design, known as "universal design," holds potential as an approach to the creation of distance learning environments. As discussed by Rose and Meyer (2002), the barriers to learning are not inherent in individuals but arise instead through learners' interactions with inflexible educational materials and methods. Basically, the principle of universal design is based on the commonsense notion that we need to make designs inclusive, useable, and accessible by as many people as possible. In commenting on universal design, Moore (2007) notes that this approach leads to "designs that incorporate greater flexibility, multiple modalities, and an understanding that we do build truly optimal instructional and performance support systems and that we do not … limit by design" (p. 534). Sims (2008) supports the need to challenge existing instructional design approaches that centralize the power of the teacher and the institution, and search for radical perspectives that capitalize on the connectivity, collaboration, and communicative potential of social networking tools. In the following sections, we consider how the affordances of social software tools can be exploited to add value to and transform teacher–learner and learner–learner interactions to better meet the needs of distance students in the new millennium.

Web 2.0 and Social Software: Affordances for Distance Education

Web 2.0 and social software tools have tremendous potential to help address or alleviate many of the aforementioned problems and barriers of distance education, including those relating to teacher contact and student support. Yet at the same time, used inappropriately and in the absence of appropriate strategies, they run the risk of further isolating and alienating distance learners, in addition to introducing technical overhead that acts as a further impediment to learning. It is therefore necessary to carefully consider the affordances of these emerging tools and technologies, as well as the dynamics of the affordances and the limitations and constraints that may be present.

An affordance is an action that an individual can potentially perform in his or her environment by using a particular tool ("Affordance," 2008); for example, blogging entails typing and editing, which are not affordances, but in tandem with other functions, lend themselves to the affordances of idea sharing and interaction. Salomon (1993) advocates analyzing information and communication technologies (ICTs) from the perspective of their educational affordances. According to Kirschner (2002), educational affordances can be defined as the relationships between the properties of an educational intervention and the characteristics of the learner that enable certain kinds of learning to take place. Conole and Dyke (2004) draw on social and educational theory to propose a taxonomy of the educational affordances of ICTs, which include the following identified themes: accessibility; speed of change;

diversity; communication and collaboration; reflection; multi-modal and non-linear learning; risk, fragility and uncertainty; immediacy; monopolization; and surveillance. They believe that the taxonomy will be useful as a "checklist" for practitioners, to assist them in making informed decisions about the use of different ICTs, and also to help increase their awareness of the properties of different tools and resources. This awareness will be beneficial as they design and develop learning activities and teaching plans.

Figure 4.1 Types of support for social learning provided by Web 2.0 (based on the work of boyd, 2003)

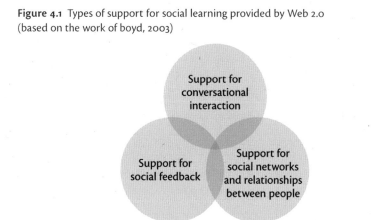

What implications do the affordances of Web 2.0 have for distance education? boyd (2003) claims that the sociability aspects of Web 2.0 have the most potential, and identifies three key, distinguishing features of social software: support for conversational interaction between individuals or groups; support for social feedback; and support for social networks and relationships between people (Figure 4.1). These overlapping elements arguably also characterize student-centered learning in distance education, and each may be viewed as an essential ingredient of social presence. Web 2.0 applications that scaffold and enable social conversation and feedback (such as blogs and wikis) create a public space for discourse and the exchange of ideas, with commentary by others, and thereby support and sustain thinking and idea creation in a community of learners. Myriad Web 2.0 tools are available that support networking and cater to the affective dimension of learning by allowing personal, engaging, and appealing interactions, and participation in networks that provide access to global communities where learners gain exposure to multiple, diverse perspectives and develop digital literacy skills.

Support for conversational interaction

At the heart of conversational theory (Laurillard, 2002; Pask, 1976) is a sense that learning and conversation are somehow linked. This idea can be connected to the theories of Vygotsky (1978), which suggest that people learn by participating in social situations, using language to share ideas and consider the ideas of others. People then internalize the ideas that are expressed in interaction, connecting these ideas into complex networks of knowledge. Though we have discovered a great deal about conversation and learning over the past several decades, we still struggle to precisely define the links between them, and the relationship of talk to reflection, concept development, and the stimulation of individual cognitive growth. Research suggests that conversation in classrooms (both face-to-face and virtual) that is less teacher-centered and more student-centered leads to improved learning. Ways of shifting the centre of attention from teacher to student include leading a discussion by asking open-ended, thought-provoking questions (Wells, 1999) and creating an atmosphere in which students feel safe enough to generate their own questions (Dillon, 1989).

Support for social feedback

The work of Lave and Wenger (1991) on learning communities and recent research on connectivism (Siemens, 2005) and communal constructivism (Tangney, FitzGibbon, Savage, Mehan, & Holmes, 2001) all emphasize the social nature of learning, and the centrality of social interaction for learning. Laurillard (2002) also mentions the importance of feedback for reflective learning, obtained through dialogic conversations that enable intrinsic and adaptive feedback processes. Laurillard believes that technology is incapable of providing customized feedback needed by learners, and emphasizes that the personalization and customization of the learning experience is essential for learning to be meaningful. Sharples' (2005) work builds on that of Laurillard, but no distinction is made between people and interactive systems such as desktops, mobiles, and ubiquitous computing devices, with the advantage that the model can be applied both to human teachers and learners, as well as to technology-based teaching and learning support systems (see also Doering, Veletsianos, & Yerasimou, 2008; Veletsianos & Miller, 2008). Nevertheless, in distance learning settings that rely on Web-based delivery, the provision of personalized feedback remains a challenge; fortunately, there are recent exemplars of designs that capitalize on the connectivity and social connectedness of innovative digital tools. Doering, Miller, and Veletsianos (2008) describe the design of adventure learning, a hybrid distance education model that enables different levels of interaction and collaboration between teachers, students, subject matter experts, and content, all occurring through the social affordances of the environment.

Support for social networks and relationships between people

Social networking sites (SNSs) and broader social networks such as the "blogosphere" allow individuals to connect, develop rapport, share interests, create community, and collaborate with peers. Many SNSs (e.g., Facebook at http://facebook.com) started among small communities of college students in the U.S., but they have now spilled over into the professional world of work (e.g., LinkedIn at http://linkedin.com). In the fields of information technology and business, these sites have multiple uses, from sales to project work, advertising and soliciting help worldwide for particular commercial ventures. Facebook connects users in multiple ways, for example, through shared virtual groups based on common interests, backgrounds, and musical tastes. Ellison, Steinfield, and Lampe (2007) maintain that the site represents a trend from offline to online relationship building as it was originally used within a bounded community. They investigated the relationship between the frequency of Facebook use by undergraduates and the development of social capital, i.e., the capacity to make and extend friendships and social acquaintances and to broaden one's worldview. In their findings, Ellison et al. report that "Facebook serves to lower the barriers to participation so that students who might otherwise shy away from initiating communication with or responding to others are encouraged to do so through Facebook's affordances" (Discussion section, paragraph 4). In education terms, building social capital means that individuals engage in collaborative activity to build knowledge, and seek support from others to solve problems — such processes are intrinsically part of lifelong learning and cognition.

Thus social software tools can also be viewed as pedagogical tools that stem from their affordances of sharing, communication, and information discovery. Some examples of the affordances of social software tools that are relevant to distance education are listed in Table 4.1. These affordances stimulate the development of a participatory culture in which there is genuine engagement and communication, and in which members feel socially connected with one another. This participation and connectivity creates a sense of community for distance education students, and provides a gateway for wider community participation and the development of essential core skills needed for lifelong learning, such as self-directed learning, knowledge creation, and digital literacy. The tools of Web 2.0 invite participation by individuals (blogs) and by groups (wikis), and learning experiences are enhanced and extended by the multiple media through which to share and communicate (e.g., podcasts, vodcasts). These tools break down the isolation of the individual and encourage active participation, thus going beyond Web 1.0-based distance education delivery, or the traditional correspondence model where users read text (or Web pages) but cannot create and contribute content, and where social interaction is limited to discussion forums, ruled by a teacher.

Exemplars of Distance Learning Pedagogies that Extend Traditional Paradigms

Table 4.2 contains a number of exemplars of how educators around the world are using Web 2.0 and social software tools to extend traditional paradigms of teaching and learning, transforming the ways in which the teacher–learner transaction (Keegan, 1996) in distance education takes place. The examples provide evidence that social computing applications such as blogs, wikis, and podcasts are catalysts for change:

Table 4.1 Examples of the affordances of social software tools

Affordance	Description and implications for distance education
Connectivity and social rapport	Social networking sites such as MySpace, Facebook, Ning, and Friendster attract and support networks of people and facilitate connections between them. They enable the creation of social capital, which refers to the capacity of people to build links and call upon others to help, test, and confirm ideas and co-create knowledge. This concept is essential to understanding why connectivity is important in distance education. The building blocks of social capital are similar to the requirements for social presence and require trust, engagement, connection, openness to others' views, and a willingness to collaborate and share ideas. Social software tools allow individuals to acquire both social and communicative skills, and at the same time become engaged in the architecture of participation of Web 2.0. Using these tools, users engage in informal learning and creative, expressive forms of behaviour and identity seeking, while developing a range of digital literacies.
Collaborative information discovery and sharing	Data sharing is enabled through a range of software applications, by means of which experts and novices alike can make their work available to the rest of the virtual world; for example, through personal and collaborative blogs. Social bookmarking tools such as Delicious, Furl, and Digg allow distance learners to build up collections of Web resources or bookmarks, classify and organize them through the use of metadata tags, and share both the bookmarks and tags with others. In this way, learners and educators with similar interests can learn from one another through subscribing to the bookmarks and tags of others, and actively contribute to the ongoing growth and evolution of the "folksonomy" of Web-based information and knowledge.
Content creation	Web 2.0 emphasizes the pre-eminence of content creation over content consumption, whereby learners can create, assemble, organize, and share content to meet their own needs and those of others. Open content initiatives and copyright models such as Creative Commons (2008) are helping fuel the growth of learner-generated content and changing the old paradigm of distance education to give learners more autonomy and scope for creativity. Wikis and other collaborative writing tools enable distributed individuals to work together to generate new knowledge through an open editing and review structure.
Knowledge and information aggregation and content modification	The large uptake of RSS, as well as related technologies such as podcasting and vodcasting, is indicative of a move to collecting material from many sources and making it available for personal needs. Hilton (2006) describes these technologies as part of a move from "producer push" to "demand pull," whereby students are now accustomed to obtaining and consuming content "on demand." There are also trends towards the unbundling of content (Hilton, 2006) and the rise of "micro-content" (Lindner, 2005, 2006; Leene, 2005), i.e., digital content in small fragments that are loosely connected and which can be "mashed-up," re-mixed, and re-formulated by individuals to produce new patterns, images, and interpretations.

the idea is to convert students into "prosumers" (producers and consumers) who create content and share it, rather than merely being passive content receivers. Using the new tools in combination with appropriate instructional designs and pedagogical strategies, students become more active learners. The real challenge for distance educators is to promote learner control, self-direction, agency, and autonomy by offering flexible options and choice, while still supplying the necessary structure and scaffolding.

Of particular significance is how the examples in Table 4.2 show that Web 2.0 and social software tools are being adopted by distance educators worldwide, and are being used in novel ways to add value to the learning process in a climate where the value of textbooks is being questioned (Moore, 2003; Fink, 2005) and where the open source and open content movements are attracting high levels of attention and application (Couros, 2006; Breck, 2007; Blackall, 2007; Schaffert & Geser, 2008). It can clearly be seen that the use of these tools presents exciting prospects for authentic learning and assessment that are directly linked to and/or situated within distance students' personal and working lives.

Table 4.2 Exemplars of Web 2.0 tools for value adding in distance learning settings

Institution and country: Charles Sturt University, Australia

Reference(s):
Lee, Eustace, Hay, & Fellows (2005); Peacock, Fellows, & Eustace (2007)

Overview of teaching/learning activity:
Distance education students undertaking a course on computer-supported collaborative work (CSCW) learn with and about collaborative groupware tools and information environments, including a range of both Web 1.0 and 2.0 technologies. The students form groups of three or four, called "PODs" (pools of online dialogue), and each group is given a fortnight to complete each of four collaborative activities/exercises.

Learner tasks:
The POD activities are not graded directly; instead, students incorporate evidence of having completed the activities, together with reflective comments on their experiences, into their individual e-portfolios, assessed at the end of the course along with other multimedia artifacts of the students' semester-

long learning journeys. Each fortnight, students are required to contribute 500 words to the class wiki; these words can be "spent" creating a new article, adding to an existing article, or pooled with other people to generate a larger article. The wiki is augmented with a five-star page rating mechanism, allowing students to rate, contribute to, and learn from one another's content.

Instructor tasks:
The instructors assist with the set-up of the technology infrastructure and develop guidelines/instructions for the fortnightly collaborative exercises, including stimulus questions to promote reflection and discussion. Instructors actively participate in PODs (as "guests") only where explicitly invited to do so by the members.

Salient pedagogical features and implications for distance education:
Through distributed, collaborative learning processes supported by social software tools, students engage in both top-down (teacher-directed) and bottom-up (learner-directed) activities, thereby enabling high levels of empowerment, freedom, and peer learning. In order to be successful, students must become actively involved in one another's learning trajectories.

Institution and country: Victoria University of Wellington, New Zealand

Reference(s):
Elgort, Smith, & Toland (2008)

Overview of teaching/learning activity:
A mixture of on-campus and distance education students undertaking a Master of Library and Information Studies program work in groups to collaboratively produce Web-based resource guides using a wiki.

Learner tasks:
Each group is required to produce three deliverables: the resource guide (a website providing links to and evaluations of information resources in a specific subject area); a presentation of the completed guide to the class; and an online reflective journal, in which students document the process of creating the guide and reflect on their personal contribution to the project.

Instructor tasks:
In this hybrid distance course, students are cast in the roles of content creators, working collaboratively and collectively to produce authentic resources for library users. The wikis and social feedback processes scaffold learner autonomy and self-regulated learning.

Salient pedagogical features and implications for distance education:
Blended or hybrid learning design that enables pedagogically supported access to resources. Group activities, interpersonal interactions, and the production/consumption of content are not controlled and constrained by the teacher, but are allowed to develop and flourish through the students' joint efforts and collective intelligence.

Institution and country: Open University, UK

Reference(s):
Kukulska-Hulme (2005)

Overview of teaching/learning activity:
Students attending German and Spanish summer schools as part of distance courses offered by the UK Open University use digital voice recorders and mini-camcorders to record interviews with other students and with native speakers of the languages they are studying, as well as to create audio-visual tours for sharing with their peers via the Web.

Learner tasks:
Learners create authentic content and tasks for peers, and in doing so have to demonstrate knowledge of and familiarity with the technology as well as genres for knowledge creation.

Instructor tasks:
The instructors supply the recording equipment and provide guidance to the students in completing the various activities; for example, by providing sample topics/questions for the student-led interviews.

Salient pedagogical features and implications for distance education:
Student activities are self-regulated and involve multiple modalities, tools, and media in various forms (e.g., text, voice, pictures). The outcomes of tasks are archived on the Web, allowing for revision and commentary by others, and this provides an example of learner-generated content.

Institution and country: University of Leicester, UK

Reference(s):
Edirisingha, Salmon, & Fothergill (2006, 2007)

Overview of teaching/learning activity:
Specialized podcasts called "profcasts" are used to enrich blended learning in

a second- and third-year undergraduate engineering module entitled Optical Fibre Communication Systems. The profcasts contain material designed to support learning distinct from that which is facilitated through structured on-campus or e-learning processes alone.

Learner tasks:
The students engage in online learning activities based on Gilly Salmon's (2002) e-tivities model. The processes are intended to add value to the learning experience and to include informal content, stimulating links to real-world applications.

Instructor tasks:
The instructor releases weekly profcasts to supplement online teaching through updated information and guidance on the weekly activities, and to motivate students by incorporating relevant news items, anecdotes, and jokes.

Salient pedagogical features and implications for distance education:
Social software is used to add value to the learning experience by enriching tasks and making them relevant and meaningful. Students' interest and motivation are increased as they are taken beyond the prescribed course content.

Institution and country: Kent State University, U.S.

Reference(s):
Byron (2005)

Overview of teaching/learning activity:
Wikis are used in a philosophy class to facilitate joint activity and the articulation of shared understanding in relation to the course content.

Learner tasks:
Each student completes various readings and posts summary reports on a wiki; the rest of the class is allowed to edit the postings to improve accuracy and completeness. The students also write five- to seven-page papers and upload them to the wiki's file gallery instead of handing in hard copies. They then engage in peer reviews of one another's papers and revise their own papers based on the feedback received.

Instructor tasks:
The instructor posts the course syllabus, schedule, and assignments on the wiki, in addition to course notes and readings. S/he also reviews the students'

summaries to obtain an indication of how well they grasped the readings. Last but not least, s/he provides a rubric to scaffold the peer review process.

Salient pedagogical features and implications for distance education: This distance learning context exemplifies the notion of the learner-generated content, and tasks are set to engage learners in peer review and commentary, thereby promoting critical thinking and reflective skills. This aligns with the social constructivist notion that knowledge must be created and validated through interaction and dialogue between individuals.

Institution and country: Aalborg University, Denmark, and Tecnológico de Monterrey, Mexico

Reference(s):
Icaza, Heredia, & Borch (2005)

Overview of teaching/learning activity:
As part of a masters-level course entitled "Culture and technology in the learning organization," Mexican students studying information technology and telecommunications and Danish students studying Spanish literature are immersed in a scenario involving employment at a virtual enterprise in the form of a fictional online publishing house that develops digital products such as e-books and Web-based tutorials.

Learner tasks:
The students assume the roles of authors hired by the publishing house. Working in teams consisting of a mixture of students from each country, they define the needs/problems that their products are to address, choose the types of products and content to develop, and finally create the products while adhering to a well-defined and stringent project methodology. Because the publishing house is structured as a learning organization that implements knowledge management strategies, the students understand that by developing their products, they are enriching the intellectual capital of the organization as well as that of future cohorts who will study the course. A wiki server is used as a repository for process documentation, reflections, and end products.

Instructor tasks:
The instructors create the simulated work environment and play the roles of editors of the publishing house. They set up a wiki to house the course Web pages representing the company intranet, comprising mission and values

statement, organizational policies, product catalogue (including e-books gener-ated in previous course offerings), job descriptions, and links to the corporate library (readings, reference materials, editing aids, etc.). The instructors also periodically review the team communication logs, providing guidance to each team in the form of questions and scaffolds rather than definitive or "correct" answers. They model the processes of reflective inquiry in the online environ-ment to encourage the students to engage in critical thinking and peer-to-peer feedback and dialogue.

Salient pedagogical features and implications for distance education: Collaborative learning, project-based learning, resource-based learning, au-thentic learning, inquiry-based learning, and learning by immersion are key pedagogical features of this example. Students are immersed in a simulated work context that engages them in a range of authentic experiences and extends their skills. This pedagogy goes beyond traditional distance educa-tion by facilitating meaningful and productive tasks that allow learners to cooperate, collaborate, and create accessible learning artifacts to be shared within the community.

Conclusion: Extending and Enriching the Student Experience in Distance Education

A large proportion of students who study online and at a distance tend to experience social isolation and technical difficulties, which may lead to de-motivation and a lack of focus in the absence of direct or regular contact with instructors and classmates (Rovai, 2002). However, Morgan and O'Reilly (1999) urge educators to view distance education from an "opportunity" model rather than a "deficit" model (p. 23), reminding them that the knowledge, skills, and abilities distance students have gained through life and work experience, as well as their access to authen-tic contexts and resources, can and should be leveraged to work to their advantage. Willis (1992) maintains that the challenges posed by distance teaching are countered and potentially outweighed "by opportunities to reach a wider student audience; to meet the needs of students who are unable to attend on-campus classes; to involve outside … [experts] who would otherwise be unavailable; and to link students from different so-cial, cultural, economic, and experiential backgrounds" (sec. 2, para. 1).

For the benefits of distance education to be fully realized, there is a need to enable maximum student self-direction, but also to simultaneously foster community building, provide individual support, and create a sense of social and teacher presence in practical, cost-effective, and sustainable ways. It has been argued that the most effective technology-supported distance learning environments are those in which social interaction is a predominant feature (Muilenburg & Berge, 2005); furthermore, both student–teacher and peer-to-peer interactions must be recognized as important factors if students are to value their studies and engage in satisfying experiences.

The social dimension of Web 2.0 tools has already begun to change the traditional paradigm of distance education, by empowering the learner and adding value to the learning experience. As Anderson (2005) states:

> It is clear that the problems that social software addresses (meeting, building community, providing mentoring and personal learning assistance, working collaboratively on projects or problems, reducing communication errors, and supporting complex group functions) have application to education use, and especially to those models that maximize individual freedom by allowing self-pacing and continuous enrolment. Educational social software (ESS) may also be used to expand, rather than constrain freedoms of their users. (p. 4)

The most important implications of social software for distance education are the new possibilities for extending and enriching the learning experience, reducing isolation, and utilizing the power and immediacy of the available tools to support the core learning processes of reflection, collaboration, knowledge creation, creativity, discussion, and social networking. The emerging functionality of the Web allows greater autonomy, freedom, and choice, but only to those who are digitally literate and capable of exploiting the tools for what they do best: support generative activity, social connectivity, and participatory learning.

It is also imperative to acknowledge that technology is intricately related to many other elements of the learning context (such as task design) that can shape the possibilities they offer to learners, and the extent to which learning outcomes can be achieved. The deployment of technologies for educational purposes must be underpinned by an explicit learning paradigm and informed by pedagogical principles that place learners at the centre of the learning process (Joyes, 2005/6; Salaberry, 2001). In response to this, the authors believe that emerging technologies — including those that are part of Web 2.0 and beyond — are best used to support and scaffold learning and reflection within authentic, real-world distance learning contexts with the aid of rich digital media (chapter 5). A range of learner-centered pedagogies, such as inquiry and problem-based learning, should afford students a true sense of agency, control, and ownership of the learning experience, together with the capacity to create, share, and communicate ideas and knowledge. To deliver the promise of quality in distance learning, we need to leverage the new tools to extend and transform current practices in appropriate ways, while keeping learners and the social dimensions of learning at the forefront (chapter 14).

REFERENCES

Affordance. (2008). In *Wikipedia, The Free Encyclopedia*. Retrieved 17 December 2008, from http://en.wikipedia.org/wiki/Affordance

Anderson, T. (2005). *Distance learning: social software's killer app?* Keynote paper presented at the 17th Biennial Conference of the Open and Distance Learning Association of Australia, Adelaide, SA, 9–11 November. Retrieved 10 August 2008, from http://www.unisa.edu.au/odlaaconference/PPDF2s/13%20odlaa%20-%20Anderson.pdf

Anderson, T. (2007). *Reducing the loneliness of the distance learner using social software*. Keynote paper presented at the 12th Cambridge International Conference on Open and Distance Learning, Cambridge, England, 25–28 September. Retrieved 3 December 2008, from http://www2.open.ac.uk/r06/conference/TerryAndersonKeynoteCambridge2007.pdf

Bennett, S., Agostinho, S., Lockyer, L., & Harper, B. (Eds.). (2009). Researching learning design in open, distance and flexible learning [Special issue]. *Distance Education, 30*(2).

Blackall, L. (2007). Open educational resources and practices. *Journal of e-Learning and Knowledge Society, 3*(2), 63–81.

boyd, s. (2003). Are you ready for social software? *Darwin Magazine*. Retrieved 8 November 2006, from http://www.darwinmag.com/read/050103/social.html

Breck, J. (Ed.). (2007). Opening educational resources [Special issue]. *Educational Technology, 47*(6).

Byron, M. (2005). Teaching with Tiki. *Teaching Philosophy, 28*(2), 108–113.

Conole, G., Culver, J., Weller, M., Williams, P., Cross, S., Clark, P., & Brasher, A. (2008). Cloudworks: Social networking for learning design. In *Hello! Where Are You in the Landscape of Educational Technology? Proceedings of the 25th ASCILITE Conference* (pp. 187–196). Melbourne, VIC: Deakin University.

Conole, G., & Dyke, M. (2004). What are the affordances of information and communication technologies? *ALT-J, Research in Learning Technology, 12*(2), 113–124.

Couros, A. (2006). Examining open (source) communities as networks of innovation: Implications for the adoption of open thinking by teachers [Unpublished doctoral dissertation]. University of Regina, Regina, SK, Canada.

Delahoussaye, M., & Zemke, R. (2001). 10 things we know for sure about learning online. *Training, 38*(9), 48–59.

Dillon, J.T. (1989). *The Practice of Questioning*. London: Routledge.

Doering, A., Miller, C., & Veletsianos, G. (2008). Adventure learning: Educational, social, and technological affordances for collaborative hybrid distance education. *Quarterly Review of Distance Education, 9*(3), 249–266.

Doering, A., Veletsianos, G., & Yerasimou, T. (2008). Conversational agents and their longitudinal affordances on communication and interaction. *Journal of Interactive Learning Research, 19*(2), 251–270.

Dron, J. (2007). Designing the undesignable: Social software and control. *Educational Technology & Society, 10*(3), 60–71.

Edirisingha, P., Salmon, G., & Fothergill, J. (2006). *Enriching a blended learning environment with podcasts and e-tivities: A case study*. Paper presented at the First International Blended Learning conference, Hatfield, England, 15 June.

Edirisingha, P., Salmon, G., & Fothergill, J. (2007). Profcasting — a pilot study and guidelines for integrating podcasts in a blended learning environment. In U. Bernath & A. Sangrà (Eds.), *Research on Competence Development in Online Distance Education and E-Learning: Selected Papers from the 4th EDEN Research Workshop in Castelldefels/Spain, 25–28 October 2006* (pp. 127–37). Oldenberg, Germany: BIS-Verlag der Carl von Ossietzky Universität Oldenburg.

Elgort, I., Smith, A.G., & Toland, J. (2008). Is wiki an effective platform for group course work? *Australasian Journal of Educational Technology, 24*(2), 195–210.

Ellison, N.B., Steinfield, C., & Lampe, C. (2007). The benefits of Facebook "friends:" Social capital and college students' use of online social network sites. *Journal of Computer-Mediated Communication, 12*(4). Retrieved 22 July 2008, from http://jcmc.indiana.edu/vol12/issue4/ellison.html

Engeström, J. (2005a, 13 April). *Object-centered sociality*. Paper presented at Reboot

7, Copenhagen, 10–11 June. Retrieved 23 July 2008, from http://aula.org/people/jyri/presentations/reboot7-jyri.pdf

Engeström, J. (2005b). Why some social network services work and others don't — Or: The case for object-centered sociality. Message posted to http://www.zengestrom.com/blog/2005/04/why_some_social.html

Fink, L. (2005, 16 September). Making textbooks worthwhile. *Chronicle of Higher Education*. Retrieved 11 March 2007, from http://chronicle.com/weekly/v52/i04/04b01201.htm

Galusha, J.M. (1997). *Barriers to learning in distance education.* Retrieved November 6, 2006, from http://www.infrastruction.com/barriers.htm

Garrison, D.R. (1997). Computer conferencing: The post-industrial age of distance education. *Open Learning, 12*(2), 3–11.

Herrington, J., Reeves, T.C., & Oliver, R. (2006). Authentic tasks online: A synergy between learner, task and technology. *Distance Education, 27*(2), 233–247.

Hilton, J. (2006). The future for higher education: Sunrise or perfect storm? *EDUCAUSE Review, 41*(2), 58–71.

Hipp, H. (1997). Women studying at a distance: What do they need to succeed? *Open Learning, 12*(2), 41–49.

Icaza, J.I., Heredia, Y., & Borch, O. (2005). Project oriented immersion learning: Method and results. In *Proceedings of the Sixth IEEE Conference on Information Technology Based Higher Education and Training* (pp. T4A/7-T4A11). Piscataway, NJ: IEEE.

Jenkins, H. (2007). *Confronting the Challenges of Participatory Culture: Media Education for the 21st Century.* Chicago: MacArthur Foundation. Retrieved 4 January 2007, from http://www.digitallearning.macfound.org/atf/cf/%7B7E45C7E0-A3E0-4B89-AC9C-E807E1B0AE4E%7D/JENKINS_WHITE_PAPER.PDF

Joyes, G. (2005/6). When pedagogy leads technology. *The International Journal of Technology, Knowledge & Society, 1*(5), 107–113.

Keegan, D. (1996). *Foundations of Distance Education* (3rd ed.). London: Routledge.

Kirschner, P.A. (2002). Can we support CSCL? Educational, social and technological affordances for learning. In P.A. Kirschner (Ed.), *Three Worlds of CSCL: Can We Support CSCL?* (pp. 7–47). Heerlen, The Netherlands: Open University of the Netherlands.

Knapper, C. (1988). Lifelong learning and distance education. *American Journal of Distance Education, 2*(1), 63–72.

Kukulska-Hulme, A. (2005). *The mobile language learner — now and in the future* [Webcast]. Plenary session delivered at the Fran Vision till Praktik (From Vision to Practice) Language Learning Symposium, Umeå, Sweden, 11–12 May. Retrieved 3 February 2006, from http://www2.humlab.umu.se/video/Praktikvision/agnes.ram

Lake, D. (1999). Reducing isolation for distance students: An online initiative. In K. Martin, N. Stanley, & N. Davison (Eds.), *Teaching in the Disciplines / Learning in Context: Proceedings of the 8th Annual Teaching and Learning Forum* (pp. 210–214). Perth: University of Western Australia.

Laurillard, D. (2002) *Rethinking University Teaching: A Conversational Framework for the Effective Use of Learning Technologies* (2nd ed.). London: RoutledgeFalmer.

Lave, J., & Wenger, E. (1991). *Situated Learning: Legitimate Peripheral Participation*. Cambridge, England: Cambridge University Press.

Lee, M.J.W., Eustace, K., Hay, L., & Fellows, G. (2005). Learning to collaborate, collaboratively: An online community building and knowledge construction approach to teaching computer-supported collaborative work at an Australian university. In M.R. Simonson & M. Crawford (Eds.), *Proceedings of the 2005 AECT International Convention* (pp. 286–306). North Miami Beach, FL: Nova Southeastern University.

Leene, A. (2005). *Microcontent is everywhere!!!* Paper presented at the Microlearning 2006 Conference, Innsbruck, Austria, 8–9 June. Retrieved 3 January 2008, from http://www.sivas.com/microcontent/articles/ML2006/MicroContent.pdf

Lindner, M. (2005, 2 September). *Wild microcontent*. Message posted to http://phaidon.philo.at/martin/archives/000318.html

Lindner, M. (2006). Use these tools, your mind will follow. Learning in immersive micromedia and microknowledge environments. In D. Whitelock & S. Wheeler (Eds.), *The Next Generation: Research Proceedings of the 13th ALT-C Conference* (pp. 41–49). Oxford, England: Association for Learning Technology.

Lockyer, L., Bennett, S., Agostinho, S., & Harper, B. (Eds.). (2008). *Handbook of Research on Learning Design and Learning Objects: Issues, Applications, and Technologies*. Hershey, PA: Information Science Reference.

Meacham, D., & Evans, D. (1989). *Distance Education: The Design of Study Materials*. Wagga Wagga, NSW: Open Learning Institute, Charles Sturt University.

Moore, J.W. (2003). Are textbooks dispensable? *Journal of Chemical Education, 80*(4), 359.

Moore, M.G. (2007). Web 2.0: Does it really matter? *The American Journal of Distance Education, 21*(4), 177–183.

Morgan, C., & O'Reilly, M. (1999). *Assessing Open and Distance Learners*. London: Kogan Page.

Muilenberg, L.Y., & Berge, Z. (2005). Student barriers to online learning: A factor analytic study. *Distance Education, 26*(1), 29–48.

Okun, M.A., Benin, M., & Brandt-Williams, A. (1996). Staying in college: Moderators of the relation between intention and institutional departure. *Journal of Higher Education, 67*(5), 577–596.

Oliver, R., Herrington, J., & Reeves, T. (2006). Creating authentic learning environments through blended-learning approaches. In C. Bonk & C. Graham (Eds.), *The Handbook of Blended Learning: Global Perspectives, Local Designs* (pp. 502–515). San Francisco: Pfeiffer.

O'Reilly, T. (2005). *What is Web 2.0: Design patterns and business models for the next generation of software.* Retrieved 15 December 2006 from http://www.oreillynet.com/pub/a/oreilly/tim/news/2005/09/30/what-is-web-20.html

Pask, G. (1976). *Conversation Theory: Applications in Education and Epistemology.* New York: Elsevier.

Peacock, T., Fellows, G., & Eustace, K. (2007). The quality and trust of wiki content in a learning community. In R. Atkinson & C. McBeath (Eds.), *ICT: Providing Choices for Learners and Learning. Proceedings of the 24th ASCILITE Conference* (pp. 822–832). Singapore: Nanyang Technological University.

Peters, O. (1992). Some observations on dropping out in distance education. *Distance Education, 13*(2), 234–269.

Race, P. (2005). *500 Tips for Open and Online Learning* (2nd ed.). New York: RoutledgeFalmer.

Rogers, P.H. (1990). Student retention and attrition in college. In R.M. Hashway (Ed.), *Handbook of Developmental Education* (pp. 305–327). New York: Praeger.

Rose, D.H., & Meyer, A. (2002). *Teaching Every Student in the Digital Age: Universal Design for Learning.* Alexandria, VA: Association for Supervision and Curriculum Development.

Rourke, L., & Anderson, T. (2002). Exploring social interaction in computer conferencing. *Journal of Interactive Learning Research, 13*(3), 257–273.

Rourke, L., Anderson, T., Archer, W., & Garrison, D.R. (1999). Assessing social presence in asynchronous, text-based computer conferences. *Journal of Distance Education, 14*(3), 51–70.

Rovai, A.P. (2002). Building sense of community at a distance. *International Review of Research in Open and Distance Learning, 3*(1), 1–16.

Salaberry, M.R. (2001). The use of technology for second language learning and teaching: A retrospective. *The Modern Language Journal, 85*(1), 39–56.

Salmon, G. (2002). *E-tivities: The Key to Active Online Learning.* London: Kogan Page.

Salomon, G. (Ed.). (1993). *Distributed Cognitions — Psychological and Educational Considerations.* Cambridge, England: Cambridge University Press.

Schaffert, S., & Geser, G. (2008). Open educational resources and practices. *eLearning Papers, 8.* Retrieved 19 October 2009 from http://www.elearningeuropa.info/out/?doc_id=13965&rsr_id=14907

Sharples, M. (2005). *Learning as conversation: Transforming education in the Mobile Age.* Paper presented at the Conference on Seeing, Understanding, Learning in the Mobile Age, Budapest, 28–30 April.

Sheets, M.F. (1992). Characteristics of adult education students and factors which determine course completion: A review. *New Horizons in Adult Education, 6*(1), 3–19.

Shih, T.K., Li, Q., & Yang, H.-C. (2007). An editorial on distance learning 2.x. *International Journal of Distance Education Technologies, 5*(3), i–iii.

Short, J., Williams, E., & Christie, B. (1976). *The Social Psychology of Telecommunications.* London: John Wiley & Sons.

Siemens, G. (2005). Connectivism: A learning theory for the digital age. *International Journal of Instructional Technology and Distance Learning, 2*(1), 3–10.

Simonson, M., Smaldino, S., Albright, M., & Zvacek, S. (Eds.). (2005). *Teaching and Learning at a Distance: Foundations of Distance Education* (3rd ed.). Upper Saddle River, NJ: Prentice Hall.

Sims, R. (2008). Rethinking e-learning: A manifesto for connected generations. *Distance Education, 29*(2). 153–164.

Suen, H.K., & Parkes, J. (1996). Challenges and opportunities in distance education evaluation. *DEOSNEWS, 6*(7). Retrieved 10 September 2006, from http://www.ed.psu.edu/acsde/deos/deosnews/deosnews6_7.asp

Tangney, B., FitzGibbon, A., Savage, T., Mehan, S., & Holmes, B. (2001). Communal constructivism: Students constructing learning for as well as with others. In C. Crawford et al. (Eds.), *Proceedings of the Society for Information Technology and Teacher Education International Conference 2001* (pp. 3114–3119). Chesapeake, VA: Association for the Advancement of Computers in Education.

Veletsianos, G., & Miller, C. (2008). Conversing with pedagogical agents: A phenomenological exploration of interacting with digital entities. *British Journal of Educational Technology, 39*(6), 969–986.

Vygotsky, L.S. (1978). *Mind in Society: The Development of Higher Psychological Processes.* Cambridge, MA: Harvard University Press.

Weller, M. (2008, 7 January). Social objects in education. Message posted to The Ed Techie website: http://nogoodreason.typepad.co.uk/no_good_reason/2008/01/whats-a-social.html

Wells, G. (1999). *Dialogic Inquiry: Toward a Sociocultural Practice and Theory of Education.* Cambridge, England: Cambridge University Press.

West, L.H.T., & Hore, T. (1989). The impact of higher education on adult students in Australia. *Higher Education, 18*(3), 341–352.

Willis, B. (1992). *Strategies for teaching at a distance. ERIC Digest.* Retrieved from ERIC database. (ED351008)

Wood, H. (1995). *Designing study materials for distance students. Occasional Papers in Distance Learning, 17. ERIC Digest.* Retrieved from ERIC database. (ED385222)

Part 2
Learning Designs for Emerging Technologies

"EMERGING":

5 A Re-conceptualization of Contemporary Technology Design and Integration

> *The Learning Technologies Collaborative,*
> *University of Minnesota*

Abstract

Within this chapter we argue that it is imperative for scholars and educators to recognize that the promise of emerging technologies is not the tool or technology itself. Instead, it is how emerging technologies are designed for and utilized in education that impacts online distance teaching and learning. As a model for our discussion, we present a theory of online learning, adventure learning as it exemplifies the power emerging technologies can have in transforming online education. Further, we detail how adventure learning as an emerging theory of online learning complements and illustrates Veletsianos' (chapter 1) definition of emerging technologies. Finally, we argue that it is the synergy between emerging technology tools, theories of online teaching and learning, and their varying affordances that will ultimately transform distance education.

Introduction

Online teaching and learning is becoming more and more widespread and ubiquitous within K–12 and post-secondary schools and institutions across the U.S. (Cavanaugh, Gillan, Kromrey Hess, & Blomeyer, 2004). Currently, there are more than three million students enrolled in completely online courses within post-secondary education, with one-third of K–12 public school districts and 9 percent of public schools (Picciano & Seaman, 2007) existing completely online or offering online courses. Florida Virtual High School served more than 31,000 students during the 2005–2006 school year. In Michigan, all current high-school students must take one online learning course before graduating. And the governor of Minnesota proposed that all Minnesota state college and university students take 25 percent of their courses online by 2015 (Bedard, 2008).

The exponential growth in online teaching and learning has fueled both the creation and use of new technologies. Additionally, "older" technologies are being built upon, and in some cases, improved with contemporary features, essentially a maturation of the original implementation ideas and design. These technologies range from courseware such as WebCT™ and Moodle™, social networking technologies such as Ning™, and Web 2.0 technologies such as wikis and blogs. Often in the numerous and diverse domains of K–12 and post-secondary education, it is inferred that if you use one of these technologies, you are "doing distance or online learning." However, for decades, scholars and researchers in the field of educational technologies have argued vehemently, and we had hoped conclusively, that it is not the media or technology that impacts teaching and learning — it is the inherent design of the technologies, their affordances, and how they are used pedagogically that facilitates successful, effective, and "good" teaching and learning (chapters 1, 2, 6, and 7; Clark, 1983; Cuban, 2009; Doering, Miller, & Veletsianos, 2008). Alone, technology tools are no more than simple media.

Cuban (2009) notes, "the real promise of technology in education lies in its potential to facilitate fundamental, qualitative changes in the nature of teaching and learning" (p. 44). Within this chapter we argue that it is imperative for scholars and educators to recognize that the promise of emerging technologies is not the technology itself. Instead, it is *how* emerging technologies are designed for and utilized in education that impacts online teaching and learning. As a model for our discussion, we discuss a theory of online learning, "adventure learning," as it exemplifies the power emerging technologies can have in transforming online education. Further, we detail how adventure learning as an emerging theory of online learning illustrates and complements Veletsianos' (chapter 1) definition of emerging technologies. Finally, we argue that it is the synergy between emerging technologies, their varying affordances, and theories of online teaching and learning that will ultimately transform distance education.

Rethinking Online Learning through Design, Curriculum, and Pedagogy

Since 2004, the adventure learning (AL) model of online learning has evolved to educate millions of students throughout the world. AL is a hybrid online learning framework that provides students and teachers with the opportunity to learn about real-world content while interacting with adventurers, students, and content experts at various locations throughout the world within an online learning environment (Doering, 2006). AL is grounded in two major theoretical approaches to learning: experiential and inquiry-based. Like experiential learning (Kolb, 1984), where learners develop understanding and meaning from their intimate experiences and reflections, within AL, students develop their understanding of subject-matter content and the world through real-time virtual experiences with teachers, adventurers, fellow students, and experts. This real-world intimate experience is the guiding goal of AL. Moreover, inquiry-based learning also guides AL, where learners are investigating the answers to their questions with little emphasis on isolated and irrelevant facts. AL uses the union of inquiry- and experiential-based learning to guide the design of its model and implementation.

Based on these theoretical foundations, the design of the adventure learning experiences follows seven interdependent principles (Doering, 2006) that further operationalize AL (see Figure 5.1):

> a researched curriculum grounded in inquiry;
> collaboration and interaction opportunities between students, experts, peers, and content;
> utilization of the Internet for curriculum and learning environment delivery;
> enhancement of curriculum with media and text from the field, delivered in a timely manner;
> synched learning opportunities with the AL curriculum;
> pedagogical guidelines of the curriculum and the online learning environment; and
> adventure-based education.

Figure 5.1 Adventure Learning model (Doering, 2006)

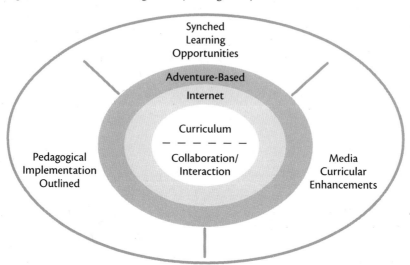

Current adventure learning projects

Reaching more than three million learners annually (across all fifty states and around the globe), previous K–12 AL programs, including the GoNorth! Series and Arctic Transect 2004 — An Educational Exploration of Nunavut, provide the grounding proof-of-concept (see http://www. polarhusky.com for current and past programs). In these programs, students across the world completed research-based lesson plans while interacting with an Arctic dogsledding expedition team, scientists, and other students and teachers. This adventure learning approach, tying existing curriculum into what is happening in society today, provides authentic and meaningful learning opportunities.

Adventure learning affordances

We believe there is a tension in the educational technology field between *what* we understand about learners and *how* we design technology-based environments that afford learning (Gaver, 1991; Kirschner et al., 2004). In other words, our understanding of learners' needs and abilities seldom reflects our awareness of the capabilities and limitations that technologies offer for instructional design. Our field tends to develop, implement, and research online and hybrid learning environments with a focus on the surface-level characteristics of the pedagogical and technological foundations of the environment (e.g., identifying optimal group sizes, performing comparative media studies, etc.), often resulting in disappointed students and instructors, diminished motivation, wasted efforts and resources, and ultimately an absence of meaningful learning (Kirschner et al., 2004). Therefore, we must re-focus our efforts not only on the technological prerequisites for meaningful learning, but also on the educational and social conditions that fuel the nature of this interaction and experience.

When designing an online learning environment, the selection and implementation of an appropriate pedagogy supportive of the instructional aims of the project, taking into account the characteristics of the selected media, is the primary concern (Kirschner et al., 2004). The social characteristics of the design must enrich the chosen pedagogy by providing engaging opportunities that encourage the social dynamics and interactions that exist habitually in traditional face-to-face learning (e.g., group formation, learner-learner and learner-instructor communication, generative problem-solving, etc.). Likewise, the technological foundation and design of the environment must not only allow for these social interactions to emerge, but ultimately thrive by providing an effective and efficient structure that satisfies users as they accomplish tasks and collaborate with peers in the environment. In this design scenario, technology is an *affordance* for learning and education, essentially a guide for the educational and social contexts of the online learning environment.

Crucial to the effective implementation of the AL model is an understanding of the *educational, social,* and *technological* affordances for delivering a successful AL project (Doering, Miller, & Veletsianos, 2008). Educational affordances are those characteristics that determine if and how effective learning takes place (Gibson, 1979; Kirschner et al., 2004; Norman, 1988), and within AL, these affordances are vital to the success of learners' experiences becoming transformational (Doering, 2006). The researched curriculum/lesson plans that accompany the online learning environment, the adventure-based approach to the AL model, and the cohesiveness of all learning activities represent the educational affordances for AL. AL social affordances are those characteristics that are instrumental in determining if and how social collaboration and interaction within the project take place. These come in the form of collaboration and interaction opportunities within the online learning environment. The technological affordances of an AL environment are (1) designed to ensure a highly usable experience for children and adult users alike, (2) scalable to an influx of both media (e.g., trail reports, photos, videos, collaboration activities, etc.) and users over the course of AL project, and (3) technology use to enhance and guide user interactions within the environment, avoiding the use of technology for technology's sake (Kirschner et al., 2004; Norman, 2004). All three of these affordances work in unison to provide opportunities for transformative learning experiences (Figure 5.2).

Figure 5.2 Affordances for AL (Doering, Miller, & Veletsianos, 2008)

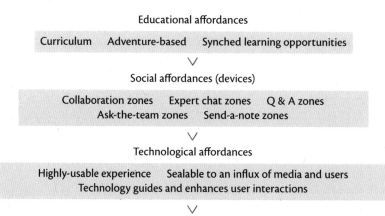

Adventure Learning Affordances

Educational affordances

Curriculum   Adventure-based   Synched learning opportunities

∨

Social affordances (devices)

Collaboration zones   Expert chat zones   Q & A zones
Ask-the-team zones   Send-a-note zones

∨

Technological affordances

Highly-usable experience   Sealable to an influx of media and users
Technology guides and enhances user interactions

∨

K–16 students, teachers, educators, experts, parents, public audience

Developing and delivering an adventure learning project

When designing and delivering the AL projects at the University of Minnesota, a region of investigation is identified (e.g., Nuanvut, Fennoscandia, Chukotka, etc.) and an inquiry-based curriculum and online learning environment is designed, developed, and delivered accordingly. For example, in preparing for GoNorth! Chukota 2007, the development of the curriculum and online learning environment focused on the region of travel—the most eastern region of Russia, Chukotka—and the four Native communities that the AL team would interact with during the expedition. The curriculum consisted of four modules that were written based on three levels of curricular activities: experience, explore, and expand (Doering, 2006).

The AL online learning environment (OLE) is developed parallel to the curriculum so the online spaces support the curricular goals and objectives. These spaces afford collaboration among learners, interaction with real-time authentic media from the field (i.e., the location of travel), delivery of authentic media that supports the curricular learning, and an overview of pedagogical principles and support for the successful teaching of AL (Doering, 2006). An example of the close connection between the OLE and the curriculum is the weekly trail reports that become available for classrooms within the OLE on Monday mornings. During the live delivery of an AL program, an "education day" is taken every Friday so the educators in the field can download the various media that was collected during the week and that support the curriculum, write and edit the text and media for the trail report, and send it to the education basecamp using satellite technologies. Once the basecamp manager receives the trail report, s/he makes sure the report and all of the activities that support the curricular goals for that week are uploaded accordingly. For example, if a curricular unit is focusing on culture, all photos, movies, QuickTime virtual reality (QTVR) files, interviews, and trail reports reinforce the culture lessons. At the same time, the education basecamp manager is updating the OLE content, scheduling the expert speaker for the week, moderating the collaboration zones where students from around the world are posting project files, and answering all questions from students and teachers to support learning and integration respectively—with all actions scaffolding the relevant curricular unit. In essence, the curricular units, media, and interactions between the actors engaged in learning (i.e., learners, teachers, explorers, and experts) support the curricular goals of the AL environment (Doering, 2006).

Adventure learning technologies

Although the goal of this chapter is not to discuss specific technologies used when designing and developing an AL program, it would be non-sensical to avoid noting the many technologies that are used to make an AL project such as GoNorth! a success. A number of diverse technologies are employed in the design and delivery of AL — some cutting-edge, some complex, others rather simple. However, it is the unison and harmony of these technologies, working both independently and together, which create opportunities for transformational teaching and learning.

To design past AL OLEs we have used diverse development technologies, including Adobe Flash or HTML/CSS for the front-end learner environment; PHP, ASP, or ColdFusion for data middleware; and MySQL or MSSQL for data storage. We have consistently challenged ourselves to think about how our design can improve the learner experience. For example, within the collaboration zones, we wanted learners to easily upload and download their project files, while at the same time being able to see other AL students from around the world with whom they were collaborating. Thus, we added an interactive map that updated in real-time with the addition of new learner-generated content (i.e., a placemark is noted on the map, along with the name of the school, student, and project file).

We also strive to ensure the OLE can be easily updated with real-time, authentic content from the field. For example, as we are travelling on the expedition route, we ensure that the curricular goals for the week guide the collection of all field data, notes, imagery, and video. We then use digital video cameras, digital cameras, microphones, handheld computers, laptops, and Iridium and Globalstar satellite technologies to make sure the weekly trail reports are available to the learner and teacher on time!

In the following sections, we situate AL within the context of emerging technologies. By discussing how AL as a practical theory of online learning is "coming into being," as well as exploring the "hype cycle" behind AL and how as a "disruptive technology" AL is not yet fully understood, we hope to illustrate AL as an emerging form/approach to online distance learning and teaching.

Adventure Learning: Conceptualizing Emerging Technologies

It is not the technology that makes AL successful; rather, it is the orchestration of numerous technologies, innovations, advancements, curricula, pedagogy, and design that makes it a success. Based on decades of discussion and debate, we have seen that it is not the AL technologies alone that make it emerging; it is the *use* of the AL model within the learning context that sets it apart and provides foundations for success.

Adventure learning is not a new technology, it is simply "coming into being"

As noted earlier in this chapter, the development technologies behind current and past AL projects (e.g., Flash, HTML, MySQL, etc.) do not represent anything new or emerging in the field; rather, the coordination and cohesive alignment of these technologies to support and make possible AL is what we consider *emerging*. Moreover, the initial AL model itself has evolved more than the technologies themselves.

With millions of students and thousands of teachers using AL on six continents, the critical question often asked at conferences, in online K–12 education discussion forums, and at speaking engagements is: "How can I create my own adventure learning program?" This is the primary inquiry we have addressed with the AL 2.0 framework, positioned at the intersection of principles, practice, and community; the often-disregarded juncture of grounded pedagogical models, practical design inquiry, and authentic context for which the framework will be implemented.

The AL 2.0 framework promises a bright future of online learning with emerging technologies, where teachers and students are delivering AL projects based on their local region of exploration and sharing their lesson goals and adventures online to collaborate with learners around the world. AL does not have to exist as an elitist form of developing learning opportunities where the region of travel is as remote as the Arctic. Rather, AL can be a class investigation of an issue or problem within the context of the learners' own locale, using the principles, practice, and community models of the AL 2.0 framework.

Through implementation of the AL 2.0 framework, we believe AL has the potential to change the existing architecture of traditional online learning by providing access to and the opportunity to collaborate and interact with authentic data, content, people, cultures, environments, and real-world contexts.

The AL 2.0 model (Figure 5.3) adds and identifies two key principles to the existing AL model: (1) the identification of an issue and respective location of exploration at the forefront of the AL project, and (2) the exploration of the issue, environment, local population, culture, and additional relevant factors that provide an authentic narrative for students and teachers to follow. These additional principles provide further support to the practical design and implementation of an AL project.

Figure 5.3 The AL 2.0 theoretical model for online learning

One of the caveats of the original AL framework was the belief that AL represents an elitist model of online education made possible only through sizeable funding and considerable development time-lines. While this may be the case in large-scale AL projects such as the GoNorth! AL Series, successful and engaging AL programs can take place in any local community over the span of a few days, even in a learner's own backyard where he or she is collecting data and media artifacts about the Mississippi River while sharing it with fellow students and teachers around the world. Therefore, by rearticulating the original framework into a practical model for integration, we encourage teachers and students to embark on their own unique AL experiences (Figure 5.4).

Figure 5.4 Practical design and implementation of an AL 2.0 project

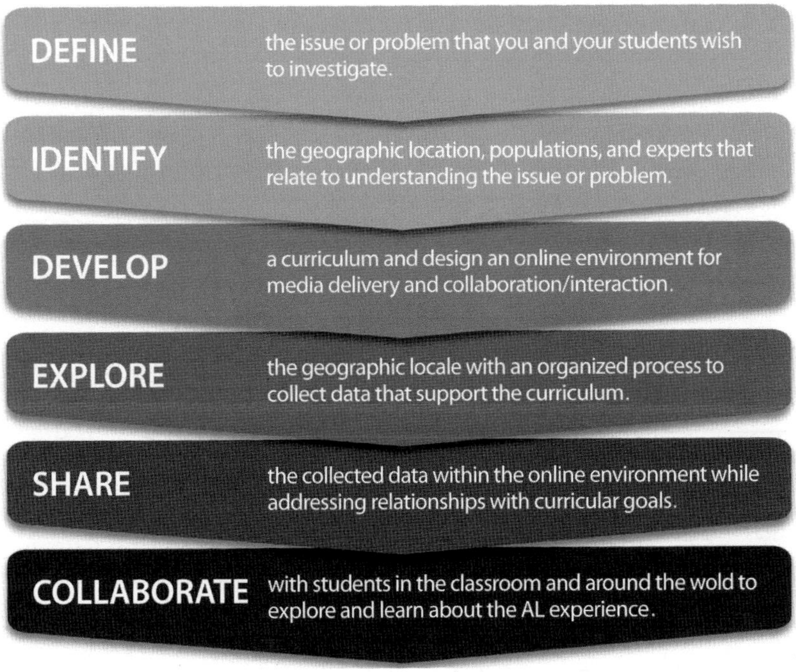

DEFINE — the issue or problem that you and your students wish to investigate.

IDENTIFY — the geographic location, populations, and experts that relate to understanding the issue or problem.

DEVELOP — a curriculum and design an online environment for media delivery and collaboration/interaction.

EXPLORE — the geographic locale with an organized process to collect data that support the curriculum.

SHARE — the collected data within the online environment while addressing relationships with curricular goals.

COLLABORATE — with students in the classroom and around the wold to explore and learn about the AL experience.

Finally, the AL 2.0 Community model (Figure 5.5) outlines the various connections and social affordances that are instrumental in determining if and how social collaboration and interaction within an AL project take place. If the synthesis of issue, place, and curriculum serves as the heart of an AL program, then collaboration and interaction would serve as the arteries and veins necessary for prolonged sustainability and vivacity. AL cannot be successful at a transformational level unless there is successful interaction and collaboration at multiple levels (chapter 2) — between students and teachers; between students and subject matter experts; between teachers and subject matter experts; between students, teachers, subject matter experts, and the AL explorers and content; and lastly, between students themselves, teachers themselves, and between the subject matter experts. The layers of interaction and collaboration occur within the social affordance devices within the project. These devices include "Collaboration Zones," "Expert Chat" zones, "Question and Answer" (Q&A) zones, "Ask the Team" zones, and "Send-a-Note" zones (see Doering, 2006; Doering, Miller, & Veletsianos, 2008).

Figure 5.5 Community collaboration in AL 2.0 (adapted from Doering, 2006)

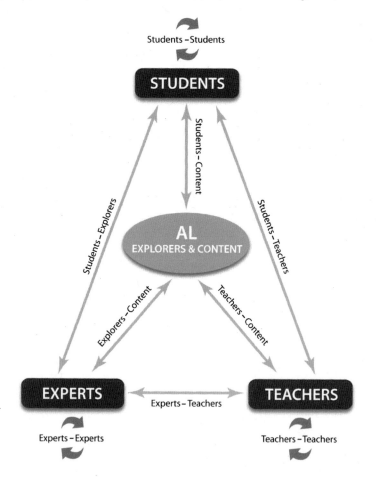

Adventure learning and the "hype cycle"

Gartner (2006) noted that technologies go through a hype cycle, and Veletsianos (see chapter 1) identifies this cycle as an indicator of an emerging technology; education theories and models also go through such cycles. For example, although not perfectly aligned, when applying this concept to the AL model, one could argue that it has gone through inflated expectations and is moving to the plateau of productivity as others implement the model for online learning. There has been much hype around AL as many individuals wish to bring attention to their favourite cause while hiking, trekking, and so on. Within six months after the completion of Arctic Transect 2004 (http://www.polarhusky. com/2004/), the University of Minnesota received dozens of calls from individuals wishing to apply the AL model to their cause for humanity — be it global climate change or how to live longer. Fortunately, because of the need and desire to do research, the need for funding to support projects, and the implementation of the model within the degree programs, AL bypassed the "trough of disillusionment" and has moved into a plateau of productivity.

Adventure learning is "not yet" fully understood

Veletsianos (chapter 1) notes that an emerging technology is not fully understood and is yet to be fully researched in a mature way. To date, many have viewed AL as being an elite, expensive approach to delivering online learning. However, this view is problematic because the AL model is not yet fully understood. The AL 2.0 evolution of the model intends to clarify this misunderstanding, as AL does not need to be an expensive undertaking. Doering (2006) noted that the AL model could be used in numerous situations if an educator wishes to give learners an authentic and real-time experience. For example, within a junior-high math class, students learn geometry through the adventure of building a home with a non-profit organization. Through connections with this "real" organization, the teacher gives students an authentic opportunity to see what using math in the real-world looks like. At a building site, students spend a day taking photos and videos, interviewing carpenters, and documenting how math is applicable outside of their classroom walls. These media artifacts are uploaded to an AL site. Parallel to the student activities, the teacher develops a lesson plan, invites an expert to answer questions related to the project, and sets up the adventure space by choosing the interactive and collaborative features that best enhance the experience. During the experience, students from around the world also share their authentic math experiences through collective artifacts online in the collaboration zones and discuss their experiences, making math real for learners. The second point is that emerging technologies are not yet fully researched or researched in a mature way. Although AL has been researched for over five years, it has by no means reached the tipping point of the masses implementing and researching the model; yet, with each new cycle of research we come to know more and more about AL and its possibilities. For example, a recent review of the adventure learning literature (Veletsianos & Kleanthous, 2009) laid the groundwork for further research in the area.

Adventure learning is "potentially disruptive"

The final characteristic of an emerging technology, or learning theory in our case, is that it is "potentially disruptive, but [its] potential is mostly unfulfilled" (chapter 1, p. 16). Projects grounded in AL can and do disrupt traditional K–12 face-to-face schooling and online teaching and learning. For example, at the completion of the first AL project, instead of attending school, hundreds of students visited the local airport to welcome the AL team back to the United States. Students have also asked their teachers why they are not using GoNorth! when other classrooms in their schools are able to participate. So, AL can pragmatically and philosophically disrupt the status quo of K–12 schooling. AL projects have also faced problems in their drive to be disruptive; although not to the extent that they could, for many reasons. First, the pedagogy inherent within adventure learning projects does lend itself more seamlessly towards constructivism. Research conducted with K–12 teachers using the adventure learning program Arctic Transect 2004 found that most participating teachers utilized constructivist learning principles in their classrooms (Doering &

Veletsianos, 2008). In addition, many teachers noted that two of the biggest hurdles to using AL programs were conflicts with standardized testing schedules and district-mandated curricula. Due to schools' ties to politics, adventure learning does not fit in the already overscheduled school calendar. Additionally, despite their hybrid nature and flexible curricula that have both on- and offline components, AL programs such as GoNorth! have yet to find their way into virtual K–12 schooling. Finally, AL seeks to disrupt barriers of digital equity and social class that often overshadow educational opportunities and experiences. Both Arctic Transect 2004 and GoNorth! were, and will continue to be, available free of charge to teachers and students: these programs require only a computer and Internet connection to access and use. Although AL has the power to disrupt and thereby transform traditional teaching, learning practices, and curricula around the world, it has yet to create a disturbance.

Designing Forward with Emerging Technologies

Imperative to the future success of our field is the investment of our collective efforts in designing and implementing "emerging" instructional solutions *with* technology. As designers, practitioners, and researchers of emerging technologies, we must challenge ourselves as a community to pose difficult design and research initiatives that bridge far-reaching gaps in technology and learning. One challenge we advocate is to place the learner experience first, with pedagogical orientation and technological selection supporting the guiding nature of the experience (see chapter 6). For example, foremost in the development and integration of an AL project, designers and teachers must strive to maintain the learner experience of excitement about learning with truly authentic content. This experience, in turn, may be achieved through designing and integrating affordances for pedagogical foundations of collaboration, inquiry-based learning, and experiential learning within the online environment. If the experience comes first, the pedagogy should fall into place.

Finally, as Reeves (2004) questioned, "Will today's passive classroom students easily transform themselves into tomorrow's active online learners?" We must explore how emerging technologies fit into this complex transformation and what roles they might play in future iterations of the learning and technology dynamic. Moreover, is it the emerging *uses* of technology, or the emerging technologies themselves, that will ultimately lead to more meaningful, transformative, and engaged online learning? Through the collaborative investigations, shared case narratives, and emerging technology research initiatives illustrated throughout this book, we believe these theoretical questions are evolving quite nicely into practical design challenges: a wonderful and welcomed progression for our field.

REFERENCES

Bedard, M. (2008) Pawlenty announces new online learning initiative during BSU visit. Retrieved 10 December 2008, from http://www.istockanalyst.com/article/viewiStockNews+articleid_2822506.html

Cavanaugh, C. & Blomeyer, R. (Eds.). (2007). *What Works in K–12 Online Learning*. International Society for Technology in Education.

Cavanaugh, C., Gillan, K., Kromrey, J., Hess, M., & Blomeyer, R. (2004). *The Effects of Distance Education on K–12 Student Outcomes: A Meta-Analysis*. Naperville, IL: Learning Point Associates. Retrieved 21 January 2007, from http://www.ncrel.org/tech/distance/k12distance.pdf

Clark, R.E. (1983). Reconsidering research on learning from media. *Review of Educational Research, 53*, 445–459.

Cuban, L. (2009). *Hugging the Middle: How Teachers Teach in the Era of Testing and Accountability*. New York, NY: Teachers College Press.

Doering, A. (2006). Adventure learning: Transformative hybrid online education. *Distance Education 27*(2), 197–215.

Doering, A., Miller, C., & Veletsianos, G. (2008). Adventure learning: Educational, social, and technological affordances for collaborative hybrid distance education. *Quarterly Review of Distance Education, 9*(1).

Doering, A., & Veletsianos, G. (2008). Hybrid online education: Identifying integration models using adventure learning. *Journal of Research on Technology in Education, 41*(1), 101–119.

Gartner Inc. (2006). *Hype Cycle for Higher E-Learning, 2006*. Retrieved 12 November 2008, from http://www.gartner.com/DisplayDocument?doc_cd=141123.

Gaver, W. (1991). Technology affordances. In S. P. Robertson, G. M. Olson & J. S. Olson (Eds.), *Proceedings of the CHI '91 conference on human factors in computing systems: Reaching through technology* (pp.79–84). New Orleans, LA: ACM Press.

Gibson, J.J. (1979) *The Ecological Approach to Visual Perception.* Boston: Houghton Mifflin.

Kirschner, P., Strijbos, J., Kreijns, K., & Beers, P.J. (2004). Designing electronic collaborative learning environments. *Educational Technology Research and Development, 52*(3), 47-66.

Kolb, D.A. (1984). *Experiential Learning: Experience as the Source of Learning and Development.* New Jersey: Prentice-Hall.

Norman, D. (1988). *The psychology of everyday things.* New York: Basic Books.

Reeves, T., Herrington, J., & Oliver, R. (2004). A development research agenda for online collaborative learning. *Educational Technology Research and Development, 52*(4), 53–65.

Veletsianos, G., & Kleanthous, I. (2009). A review of adventure learning. *The International Review of Research in Open and Distance Learning, 10*(6), 84–105. Retrieved 21 April 2010, from http://www.irrodl.org/index.php/irrodl/article/view/755/1435

DEVELOPING PERSONAL LEARNING NETWORKS FOR OPEN AND SOCIAL LEARNING

6

> *Alec Couros*

Abstract

In 2008, an open access, graduate level, educational technology course was offered at the Faculty of Education, University of Regina. The development and facilitation of this course was inspired by philosophies of the open source movement, recent trends in social media, and pedagogies designed to inspire the open, transparent, and networked learning of its participants. The outcome of this course could hardly have been anticipated. By the end of the semester, non-registered participants outnumbered registered students 10 to 1 as a larger educational community formed around the course. The resulting experience has provided insight into the potential for leveraging personal learning networks in open access and distance education.

Introduction

In January 2008, I led an open access, graduate level, educational technology course at the University of Regina titled "Education, Curriculum, and Instruction (EC&I) 831: Open, Connected, Social." This fully online course was developed and facilitated using primarily free and open source software (FOSS) or freely available services. Additionally, the course demonstrated open teaching methodologies: educational practice inspired by the open source movement, complementary learning theory, and emerging theories of knowledge. The course challenged typical boundaries common to more traditional distance education courses as students built personal learning networks

(PLNs) to collaboratively explore, negotiate, and develop authentic and sustainable knowledge networks. This latter focus became a catalyst that, as one student described emphatically, "blew the doors of this course right off their hinges." As a result, the context for learning shifted from the potentially mundane to an engaging series of events where the twenty registered students freely interacted with at least two hundred other educators, theorists, and students from around the world.

EC&I 831 has received considerable attention by academic researchers and educational bloggers. Dave Cormier (2008) wrote that the course provides "an ideal example of the role social learning and negotiation can play in learning." Jeffrey Young (2008) listed the course as one of three examples of a "growing movement" towards experimenting with open teaching in higher education. George Siemens (2008) described the design of the course as "an important source of insight" that served to inspire the development of the "Connectivism and Connective Knowledge" (CCK08) course, the inaugural Massive Open Online Course (MOOC) facilitated by Siemens and Downes. Personally, my experiences in developing and facilitating this course have been the most exciting teaching and learning experiences of my academic career. It is my hope in writing this chapter that I capture and document relevant reflections and activities to provide starting points for those considering open teaching as educational innovation.

This chapter is broken into three sections. In the first, I briefly outline key theoretical foundations that influenced the design and development of the course. This section combines philosophical, pedagogical, and practical considerations to inform a model for open teaching. In the second section, I describe the course experience in detail. This discussion includes an overview of emerging technologies used in the course and an outline of the various course activities and assessments. The third section provides discoveries related to the role of personal learning networks, outlines techniques for developing and leveraging PLNs in distance education courses, and describes the role of emerging technologies in building and facilitating networked interactions.

Theoretical Foundations

Several overlapping bodies of theory and practice informed the development and facilitation of EC&I 831. This section briefly identifies relevant points from the following areas: the open movement, complementary learning theories, and connectivism. The section ends with a description of how these areas informed a model of open teaching for the course.

The open movement

In 2003, I initiated a two-year-long study that examined the perceptions, beliefs, and practices of educators who participated in free and open source software (FOSS) communities (Couros, 2006). Through data collection and analysis, it was revealed that the majority of participants were strongly influenced by the dominant philosophical views inherent within these FOSS communities. Participants identified strong tendencies towards collaboration, sharing, and openness in their classroom activities and through professional collaborations. Generally, these individuals identified themselves as part of a larger phenomenon, later defined as the "the open movement."

> The open movement is an informal, worldwide phenomenon characterized by the tendency of individuals and groups to work, collaborate and publish in ways that favour accessibility, sharing, transparency and interoperability. Advocates of openness value the democratization of knowledge construction and dissemination, and are critical of knowledge controlling structures. (Couros, 2006)

In the early stages of this study, participants expressed frustration with perceived barriers that limited the adoption of openness in their practice. Several technical barriers were identified (e.g., software not available, suitable, or mature; sparsely available content), but soon, many of these issues improved or were resolved. One of the most advantageous developments was perceived to be the sudden popularization and availability of Web 2.0 tools. Study participants and their students alike had now gained the ability to *easily* create, share, and collaborate through emerging technologies such as blogs, wikis, podcasts, and social networks. Coinciding with this greater access to publishing came the greater availability of educationally relevant content. Participants gained access to information resources such as Wikipedia, course content through initiatives such as MIT OpenCourseWare and the OER Commons, and multimedia and video content through services such as YouTube. The dilemma of the educator shifted quickly from a perceived lack of choice and accessibility to having to acquire the skills necessary to choose wisely from increased options.

Other relevant discoveries from this study included differences in the practical and philosophical beliefs of participants. The positioning of each individual ranged from open source zealot to hobbyist; from those who refused to use *any* proprietary software, to others who voiced more practical beliefs regarding the adoption of tools. To a FOSS purist, the perceptions of the latter group would likely be considered unacceptable. For the professional educator, these more practical beliefs supported greater options for the adoption of emerging technologies. It is this latter, more general, view of openness that informs my emerging framework for open teaching.

Complementary learning theories

Several learning theories have influenced my approach to distance education and online learning. These include social cognitive theory, social constructivism, and adult learning theory (andragogy). As much has been written regarding each of these theories, this section serves only to highlight key points of each theory as it relates to the emerging concept of open teaching.

Social cognitive theory (SCT), also known as social learning theory, suggests that a combination of behavioural, cognitive, and environmental factors influences human behaviour. SCT posits that humans learn through their observations of other individuals. If one observes particular behaviours that become associated with favourable outcomes, such behaviours are more likely to be adopted by the observer (Albert & Bandura, 1963). Another relevant feature of SCT is Bandura's (1997) concept of self-efficacy that he defines as "people's judgment of their capabilities to organize and execute courses of action required to attain designated types of performances" (p. 391). Bandura considered self-efficacy beliefs to be the most influential arbiter of human activity and an important element in conceptualizing student-centred learning environments (Lorsbach, 1999).

The theory of social constructivism, attributed to Vygotsky, is related to social cognitive theory in that both theories emphasize the importance of the sociocultural context and the role of social interaction in the construction of knowledge (Woolfolk & Hoy, 2002; Derry, 1999). Instructional models influenced by social constructivist perspectives highlight the importance of collaboration among learners and practitioners in educational environments (Lave & Wenger, 1991). Another important feature of social constructivism is the concept of the zone of proximal development (ZPD). The ZPD is commonly expressed as the difference between what a learner can do independently and what the same learner can do when tutored (Vygotsky, 1978). Moving beyond tutoring, Tabak (2004) introduced the concept of distributed scaffolding, an emerging approach of learning design that incorporates multiple forms of support that respond to the diversity of learner needs and to the complexity of given learning environments. Through a greater understanding of how individuals construct knowledge and skills, the role of the social environment, and the design of flexible learner support, educators can increase student performance in both face-to-face and distance learning environments.

Adult learning theory, also known as andragogy, is based on the perception that adults learn differently than children, and that these differences should be acknowledged and accommodated. Knowles, primary developer of this theory, argued that adults generally possess different motivations for learning and have acquired significant life experiences; both of these factors greatly influence the learning process (1970). Due to these key differences, Knowles proposed the following principles for adult learning:

(1) Adults need to be involved in the planning and evaluation of their instruction.

(2) Experience (including mistakes) provides the basis for learning activities.

(3) Adults are most interested in learning subjects that have immediate relevance to their job or personal life.

(4) Adult learning is problem-centred rather than content-oriented. (p. 43)

These general principles proved to be beneficial in supporting the learning of the participants of EC&I 831.

Connectivism

Connectivism, originally developed by George Siemens (2004), is a "Net aware" theory of learning and knowledge (see chapter 2, this volume) that is heavily influenced by theories of social constructivism (Vygotsky, 1978), network theory (Barabási, 2002; Watts, 2004), and chaos theory (Gleick, 1987). Connectivism emphasizes the importance of digital appliances, hardware, software, and network connections in human learning. The theory stresses the development of "metaskills" for evaluating and managing information and network connections, and notes the importance of pattern recognition as a learning strategy. Connectivists recognize the influences that emerging technologies have on human cognition, and theorize that technology is reshaping the ways that humans create, store, and distribute knowledge.

The following principles of connectivism were most relevant to the development and facilitation of EC&I 831:

> Learning and knowledge rests in diversity.
> Dynamic learning is a process of connecting "specialized nodes" (people or groups), ideas, information, and digital interfaces.
> "Capacity to know more is more critical than what is currently known."
> Fostering and maintaining connections is critical to knowledge generation.
> A multidisciplinary, multi-literacy approach to knowledge generation is a core to human learning.

> Decision-making is both action and learning: "Choosing what to learn and the meaning of incoming information is seen through the lens of a shifting reality." (Adapted from Siemens, 2005)

A connectivist approach to course design acknowledges the complexities of learning in the digital age. The theory offers insight into how learning can be managed through the better understanding of emerging technologies and their relationship to knowledge networks.

Open teaching

Through an exploration of the above influences, I developed the following definition for the concept of open teaching. This definition helped to inform the epistemological, philosophical, and pedagogical considerations for EC&I 831.

Open teaching is described as the facilitation of learning experiences that are open, transparent, collaborative, and social. Open teachers are advocates of a free and open knowledge society, and support their students in the critical consumption, production, connection, and synthesis of knowledge through the shared development of learning networks. Typical activities of open teachers may include some or all of the following:

> advocacy and use of free and/or open source tools and software wherever possible and beneficial to student learning;
> integration of free and open content and media in teaching and learning;
> promotion of copyleft content licenses for student content production and publication;
> facilitation of student understanding regarding copyright law (e.g., fair use/fair dealing, copyleft/copyright);
> facilitation and scaffolding of student personal learning networks for collaborative and sustained learning;
> development of learning environments that are reflective, responsive, student-centred, and that incorporate a diverse array of instructional and learning strategies;

> modelling of openness, transparency, connectedness, and responsible copyright/copyleft use and licensing; and,
> advocacy for the participation and development of collaborative gift cultures in education and society.

Open teaching is an emerging concept, and this most current framework is one that guided the design of EC&I 831.

EC&I 831 in detail

This section will provide thorough detail of the development and facilitation of EC&I 831. Covered areas include a general overview of the course, details of the project's initiation, arguments for the primary learning environment, and a description of the course facilitation model.

Overview of the Course

EC&I 831 is a graduate studies in education course that focuses on the appropriate and critical integration of technology and media in K–12 classrooms. The course is not new — it has been around since 2001 — but when originally submitted to the university calendar, it was written broadly enough to provide sufficient flexibility for future course development. This feature has allowed for its extensive tailorability and responsiveness to changes in the field of educational technology, from the shifting focus (e.g., recently from eLearning to social learning) to the types of emerging technologies available to universities and colleges.

The section of the course discussed in this chapter ran from January to April 2008. There were twenty registered students, most of whom were practicing teachers (K–12) or educational administrators. The graduate courses in our faculty have a typical maximum of sixteen students, but I requested an overload due to student interest in the course and because of the peer-supported pedagogical approach proposed.

Project initiation

The Government of Saskatchewan offers Technology Enhanced Learning grants for the development of online courses, and $30,000 was awarded for EC&I 831. Typically, when granted an award, the "content expert" (myself in this case) is assigned instructional design and multimedia support personnel. For EC&I 831, I opted out of this support for three main reasons. First, I possess a strong background in instructional design and multimedia. While the university support for these areas is excellent, in envisioning the design of *this* course, I did not feel these were the areas on which I wanted to spend the bulk of the grant money. Second, considering the type of course I was teaching, I felt that there was no better way to research the area of emerging technologies than to immerse myself in the design, development, and testing of the various tools and strategies. These activities were powerful in ascertaining the various advantages, disadvantages, and social affordances inherent within the various tools implemented. This flexibility also avoided being locked into a tool that did not pedagogically or practically suit the emerging needs of learners. Finally, I identified that the area of support most needed for this course was in the development and support of the participants' personal learning networks. Thus, two learning assistants were hired as social connectors, and their primary responsibilities were to support students in the development of PLNs. These connectors were not tied to a tool or to a learning environment, but directly to the participants — their technical experience, their unique needs for support, and their learning goals.

Primary learning environment

In the weeks preceding the course, there was much research and discussion regarding the choice of a primary learning environment. Several were tested, and the following gives a brief overview of our conclusions.

WebCT (now BlackBoard) At the time of the course, WebCT was the officially supported Course Management System (CMS) at the University of Regina. WebCT was appealing for two reasons: the university had a strong infrastructure of support for WebCT, and the enrolled students had prior experience with the environment. WebCT was rejected, however, for the following reasons: it was a proprietary system that could not be modified without vendor support, the learning environment favours directed learning rather than constructivist approaches, and licensing fees were expensive and increasing in cost. Additionally, a goal of EC&I 831 was that students would be able to explore tools in the course and then apply them to their own professional work. WebCT was not freely available, free, or low-cost, and participants would not likely have much access to this tool in their school divisions.

Moodle Most readers will know that Moodle is a free and open source Course Management System that has been adopted with success in a number of educational institutions. Moodle was seen to be more favourable than WebCT in that Moodle is FOSS, modifiable, and has strong community support. Moodle also touts a "constructivist and social constructionist" approach to learning through its full range of tools and modularity. Moodle is also more available to course participants, although the software requires particular server infrastructure (e.g., PHP) and technical expertise, leading to hidden costs such as hiring a developer to setup the platform (chapter 10). The reasons we did not choose to adopt Moodle include: the software was not as easily available to participants as we hoped, the concept of the CMS is heavily course-centric rather than student-centric, and the majority of Moodle content modules represent a top-down instructivist approach to learning.

Ning Ning is an online platform that allows users to create their own social networks. Ning is not considered a CMS tool, but because social networking was to be an important activity in EC&I 831, Ning was a strong candidate for a primary learning environment. Ning's favourable characteristics include: ease of use, freely available, familiar functionality for Facebook users, community- and individual-level privacy options, user-centric spaces, content aggregation, and the inclusion of basic communication tools. The reasons we did not choose to adopt Ning include the lack of a wiki feature and the awkwardness in including core content material (e.g., syllabus, scope-and-sequence, assessments).

Wiki A Wikispaces.com hosted wiki was the primary environment chosen for EC&I 831. We reviewed several FOSS wiki software engines (MediaWiki, MindTouch Deki, TikiWiki, PhPWiki), as well as three hosted wiki services (Wikispaces, PBWiki, WetPaint). While we desired the level of administrative and data control a self-hosted option would give us, we were hesitant due to the time cost identified for patches, updates, and spamming issues. A hosted service provided us with strong technical support, and we could avoid advertising for a small monthly fee. We chose Wikispaces.com as it was the senior, best-known, and most stable of the three major providers, offered solid technical support, allowed options for CSS/theme modification, and had a simple user interface that supported many third-party services. The resulting wiki can be found at http://eci831.wikispaces.com.

Course facilitation model

The following section will outline and describe the course facilitation model through a description of the major assessments and related activities performed by course participants.

Major assessments Three major student assessments guided the activities of participants for EC&I 831: the development of a personal blog/digital portfolio, the collaborative development of an educational technology wiki resource, and the completion of a student-chosen, major digital project. Activities related to each of these assessments were designed to require and/or result in the development of a personal learning network. Thus, PLNs were both the prerequisite to and the outcome of successful completion.

(a) *Personal Blog/Digital Portfolio:* Each participant was responsible for developing a digital space to document his or her learning through readings and activities, to provide a space for personal reflection, and to create a personal hub for networked connections. In most cases, these spaces quickly became showcases of student professional activity, and acted as distributed communication portals — alternatives to centralized, managed discussion forums. Students chose from a number of free services to host their spaces (e.g., Wordpress.com, Edublogs.org) and each blog was customized by the user, both functionally and aesthetically. In most cases, these blogs continue to be maintained and have remained active well beyond the official end date of the course.

(b) *Collaborative Wiki Resource:* Students worked collaboratively to develop the content of a wiki focused on the use of technology in education. The resource, found at http://t4tl.wikispaces.com, is the result of hundreds of student edits, and covers topics such as tools and techniques, digital pedagogies, virtual worlds, mobile learning, course management software, digital storytelling, podcasting, and screencasting. The site also provides case studies of technology use in the classroom that are supported by rich, multimedia examples.

(c) *Major Digital Project:* The major digital project was designed so that students could develop a relevant resource for their specific professional context. Some students produced videos, instructional resources, or other multimedia. Others engaged in social networking activities: participation in global collaborative projects, development of private social networks, and development of localized professional development workshops. The completed activities represented a vast range of student technological competencies as well as professional and personal interests.

Tools and Interaction There were a number of synchronous and asynchronous interactions designed throughout the course. This section outlines these interactions and describes the tools used.

(a) *Synchronous Activities:* Two synchronous events were planned weekly, and these averaged in length from 1.5 to 2 hours. The first session of the week was focused on content knowledge and in connecting students to leaders in the educational technology community. Ten presenters in all were invited, and these included Canadian educational leaders such as Dr. Richard Schwier, George Siemens, and Stephen Downes. All sessions were interactive and recorded in various formats, including an audio-only podcast version. The second session of each week was a "hands-on" session where participants would learn both technical skills related to the dozens of tools used in the course, as well as the tools' pedagogical possibilities.

Several tools were used to facilitate the synchronous sessions. Adobe Connect, a proprietary web-conferencing tool, was first chosen as a relatively inexpensive solution. Unfortunately, Connect was dropped after only two sessions as we experienced poor audio, system crashes, and negative user feedback. Elluminate, a more expensive alternative, was used next. This tool was found to be more stable, but students and presenters complained about the "primitive" user interface and system crashes. The larger identified issue was that the tool was expensive, proprietary, and not available to most of the participants for their own use. Finally, we began to experiment with ustream.tv (a free video-streaming service) in combination with Skype audio-conferencing. The combination of these two free services created a stable video-conferencing tool that became the preferred choice for course participants and presenters. More importantly, unlike both Connect and Elluminate, this configuration was not bound by a licensed seat limit. This allowed us to invite other "informal" participants from outside the official course. A precise description of how ustream.tv and Skype were used can be found at http://educationaltechnology.ca/couros/765.

(b) *Asynchronous Activities:* Participants also engaged in a number of asynchronous activities between our weekly sessions. Some of the most common activities of participants included:

> reading, reviewing, and critiquing course readings through participant blogs;

> sharing and reviewing articles, tools, and readings through participant blogs or through posting to Delicious (social bookmarking service) with the common course tag (i.e., eci831readings);

> creation of screencasts, tutorials, or other resources for self-referencing or to assist other participants' understanding;

> reading, reviewing, commenting, and subscribing to blogs from outside of the course community;
> participation in open, viral professional development opportunities (e.g., Edtech Talk, OpenPD);
> posting created content to Youtube, Blip.tv, ustream.tv, Diigo, Voicethread, Mind42, Google Docs, or other collaborative, social media services;
> microblogging through Twitter or Plurk;
> collaborative design and development of lesson plans or instructional sets; and,
> continued development of the collaborative course wiki.

Many of the asynchronous activities were completely unplanned. Participants worked with individuals in the course community, but often, strong bonds formed with individuals outside of the course due to common interests. Through both the synchronous and asynchronous activities, personal learning networks grew as individuals freely connected with those interested in the content and collaboration, and not solely because of the identification with a specific course. Social interactions became authentic, dynamic, and fluid.

Personal Learning Networks in Distance Education

The first synchronous session of EC&I 831 was a private session with only the registered course participants in attendance. In this session, I briefed students about the potentially open nature of this course and that non-registered participants would be brought in to give formal presentations, to comment on student blogs, and to interact in other ways not yet known. Although optimistic, I was not yet sure at the time how I would solicit interaction from "outsiders" with these students. Yet, only two to three weeks into the course, it became evident how important the development and utilization of my PLN would be in supporting the pedagogical model of the course. To share these understandings, this section will provide a brief definition of personal learning networks and provide strategies for leveraging PLNs in distance education courses.

Conceptualizing the PLN

When I began conceptualizing this chapter, I envisioned a literature review focused on the differences between personal learning environments (PLEs) and personal learning networks. While there is a growing field of research and thinking behind the concept of the PLE (chapter 9), the academic research on PLNs is much more anecdotal. A quick Google search will deliver hundreds of blog entries highlighting the importance of the PLN, dozens of strategies focused on how to build a PLN, and many K–12 conference presentations focused on the PLN as professional development. Yet, a definition of the PLN — one that differentiates itself from the PLE — does not readily exist.

Long before I read anything about the PLN, I discovered a variation of the concept as it emerged in the practice of the participants of my doctoral study. Through this research, I noted a significant increase in the social connectivity related to the practice of study participants. This phenomenon was a vast departure from what was understood as a "typical teacher network," one often bound by local curriculum, school district, and geography. I developed two diagrams (Figure 6.1 and Figure 6.2) informed by the aggregate data, which describe the differences in the two networks.

Figure 6.1 Typical teacher network (from Couros, 2006)

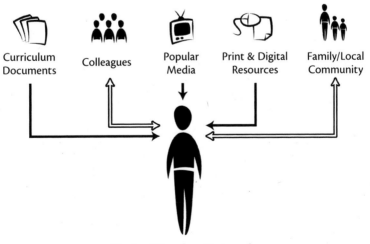

Curriculum Documents   Colleagues   Popular Media   Print & Digital Resources   Family/Local Community

Typical Teacher Network

Figure 6.2 The networked teacher (from Couros, 2006)

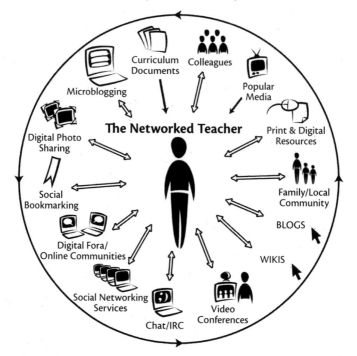

I consider "the networked teacher" representation to be a PLE diagram. It describes an individual's connectivity through participation in social media activities (e.g., blogging, wikis, social networking), and the arrows represent both the consumption and production of content.

In seeking a distinction between the PLE and PLN, I performed a recursive exercise. I asked individuals within what I perceive to be my own PLN about their perceptions of the differences between the two terms. This interrogation was facilitated via Twitter, as this microblogging platform has proven for me to be the most responsive method for surveying the connective knowledge of those within my PLN. The result was a steady outpouring of educators who offered definitions, print and multimedia resources, and diagrams, many of these developed personally or by those to whom they were connected. I have attempted to capture this conversation in a blog post at: http://educationaltechnology.ca/couros/1156.

The general consensus of this conversation maintains that PLEs are the tools, artefacts, processes, and physical connections that allow learners to control and manage their learning. This definition supports Martindale's and Dowdy's observation that "A PLE can be seen as a manifestation of a learner's informal learning processes via the Web" (chapter 9). Definitions of PLNs, however, seem to extend this framework to more explicitly include the human connections that are mediated through the PLE. In this framework, PLEs become a subset of the substantially humanized PLN. For reference in the remainder of this section, my PLN definition is simple: personal learning networks are the sum of all social capital and connections that result in the development and facilitation of a personal learning environment.

PLNs for teaching and learning

The following is a short list of strategies for developing a personal learning network and for leveraging the PLN in distance education courses. These points were effective in the facilitation of EC&I 831 as evidenced by personal reflection and student feedback.

Immerse Yourself. The entire PLN strategy depends on the use and understanding of social media in the formation of human networks. The essential tools in my own experience are blogging (self-hosted Wordpress), wikis (Wikispaces), social bookmarking (Delicious), photo sharing (Flickr), video sharing (Youtube, BlipTV), and microblogging (Twitter, Plurk). Understanding how these tools work, how they can be used together, and how your students can utilize them is essential.

Learn to Read Social Media. Although the situation is improving, traditional search engines are not currently ideal for reading social media. There are a number of social media search engines and tools available that are important to understand. Specialized search tools such as Technorati, Google Blog Search, or WhosTalkin allow for better search results. Social media browsers such as Flock and tools such as Feedly have been developed for those who primarily view, produce, and interact with social media. Blog aggregators such as Google Reader or Bloglines are essential for tracking student work if blogging is assigned. Social media is read much differently than traditional media.

Strengthen Your PLN. Human connections in PLNs are strengthened through various degrees and forms of interaction. Producing content is an important activity that may include writing and sharing blog posts, media, content, and links to resources. Feedback on the contribution of others is also equally important for social bonding and bridging. Providing comments on media, participating in digital conferences, or contributing to community resources strengthens your PLN.

Know Your Connections. Through continuous interaction, I was able to form a strong comprehension of the backgrounds and skills of many of the individuals within my PLN. This was of great benefit for me as I was then able to refer my students to educators who I knew would be willing to assist and provide expertise in areas where I lacked knowledge or experience. These interactions would often benefit learners in the extension of their own PLN.

PLNs Central To Learning. The most transformative realization that occurred to me because of EC&I 831 is just how important PLNs are for sustained, long-term learning, for students and facilitators alike. Having taught dozens of courses through CMS tools, I think of the irony: the tremendous amount of time and effort put into the development of local, time-based, course-centric communities. The communities die, usually only days after the official end-of-course date. They die because they are communities based around courses, not communities based around communal learning. For students who developed PLNs in EC&I 831, their learning communities still exist. The individuals are active and interactive, and continue to form and negotiate the connections they need to sustain long-term learning for themselves and for their students. This will be further evidenced when the EC&I 831 Class of '08 visits the current student cohort this semester — an idea for collaboration initiated by these former students.

Final Thoughts

I have given several conference presentations based on my experiences with EC&I 831. The two most commonly asked questions from audience members are phrased similarly to "How did you get away with this?" and "Where do you find the time to teach this way?" In answer to the first question, I cannot overemphasize the importance of institutional support for open teaching. I consider myself lucky to work within a faculty of education whose members I would characterize as constructively critical of technology, but strongly supportive of innovation in teaching and learning. Additionally, social justice is an integral theme in our faculty programming, and open teaching supports similar philosophies and the need for more accessible learning in our communities and in our greater society. To the second question, my gut response is to note my personal belief that *good* teaching always requires more time. This response is often not well received, considering the "publish or perish" mantra evident in contemporary universities. I can only support this with my own experience, through the realization of how interlaced the activities of teaching, learning, and research have become through the development of my own personal learning network. When I contribute to the network, I am rewarded with potentially rich opportunities for student learning, connections to individual knowledge and expertise, and tremendous insight into emerging areas of research. While developing a PLN requires a significant time commitment initially, these losses can be regained quickly through networked efficiencies, enhanced learning experiences, and new opportunities.

In summary, this chapter highlighted some of the key processes involved in the development and facilitation of EC&I 831. Careful attention to the course's theoretical foundations, use of emerging technologies, and personal network building assured the success of this course for its students. Perhaps the most telling quote regarding the success of the course comes from Jennifer, who wrote, "The best part of this course is that it's not ending. With the connections we've built, it never has to end."

REFERENCES

Albert., & Bandura, R.H. (1963). *Social Learning and Personality Development.* Austin: Holt, Rinehart & Winston.

Bandura, A. (1997). *Self-Efficacy: The Exercise of Control.* New York, NY: Worth Publishers.

Barabási, A. (2002). *Linked.* New York: Penguin Group USA.

Cormier, D. (2008). Rhizomatic education: Community as curriculum. *Innovate, 4*(5). Retrieved 27 May 2008, from http://www.innovateonline.info/index.php?view=article&id=550

Couros, A. (2006). Examining open (source) communities as networks of innovation: Implications for the adoption of open thinking by teachers [Unpublished doctoral dissertation]. University of Regina, Regina, SK, Canada.

Derry, S.J. (1999). A fish called peer learning: Searching for common themes. In A. O'Donnell & A. King (Eds.), *Cognitive Perspectives on Peer Learning* (pp. 197–211). Mahwah, NJ: Lawrence Erlbaum.

Gleick, J. (2002). *What just happened: A chronicle from the learning frontier.* London: Fourth Estate.

Knowles, M. (1970). *The Modern Practice of Adult Education: Andragogy versus Pedagogy.* Washington, DC: Association Press.

Lave, J., & Wenger, E. (1991). *Situated Learning: Legitimate Peripheral Participation (Learning in Doing: Social, Cognitive and Computational Perspectives).* New York, NY: Cambridge University Press.

Lorsbach, A. (1999). Self-efficacy theory and learning environment research. *Learning Environments Research, 2*(2), 157–167.

Siemens, G. (2004). *Connectivism: A learning theory for the digital age.* Retrieved 5 January 2008 from www.elearnspace.org/Articles/connectivism.htm

Siemens, G. (2008). On finding inspiration. Retrieved 30 June 2008, from http://ltc.umanitoba.ca/connectivism/?p=25

Tabak, I., (2004). Synergy: A complement to emerging patterns of distributed scaffolding. *The Journal of the Learning Sciences, 13*(3), 305–335.

Vygotsky, L. (1978). *Mind in Society: Development of Higher Psychological Processes.* Cambridge: Harvard University Press.

Watts, D.J. (2004). *Six degrees: The science of a connected age.* New York: W.W. Norton & Company.

Woolfolk, A., & Hoy, W. (2002). *Instructional Leadership: A Learning-Centered Guide.* Boston: Allyn & Bacon.

Young, J. (2008, 26 September). More "open teaching" courses, and what they could mean for colleges. *The Chronicle of Higher Education.* Retrieved 8 January 2009, from http://chronicle.com/wiredcampus/article/3349/more-open-teaching-courses-and-what-they-could-mean-for-colleges

CREATING A CULTURE OF COMMUNITY IN THE ONLINE CLASSROOM USING ARTISTIC PEDAGOGICAL TECHNOLOGIES

7

> *Beth Perry & Margaret Edwards*

Abstract

Those who are interested in emerging practical educational strategies for facilitating effective online instruction will find this chapter informative. Exemplary online educators employ teaching technologies that optimize meaningful interaction, facilitating an ongoing social experience that helps create a culture of community (Perry & Edwards, 2005, 2010). Many of these strategies that help establish an online educational culture of community share one aspect: they are founded in the artistic. That is, they include literary, visual, musical, or dramatic elements. We have labelled these "artistic pedagogical technologies" (APTs). In this chapter, APTs are defined, and examples are provided. Current related literature is summarized. Explanations regarding how APTs encourage interaction, create social presence, and facilitate a culture of community in the online educational milieu are proposed. Vygotsky's (1978) Social Development Theory (SDT) frames the discussion. Philosophical, theoretical, and pedagogical shifts that could influence the development, adoption, and use of APTs are described.

Introduction

Advances in Internet-based technology have changed the social and pedagogical perspectives of online learning (Dabbagh, 2004). Many online educators have moved philosophically from objectivism to constructivism, theoretically from behaviourism to socio-cognitive views of education, and pedagogically from supporting direct instruction to championing collaborative learning (Shea, 2006). In step with these foundational shifts, scholars emphasize values such as interaction, social presence, and community in the post-secondary online classroom (Dabbagh, 2004; Hodge et al., 2006; Rourke, Garrison, Anderson, & Archer, 2000; Shu-Fang & Aust, 2008). There is limited research exploring specific teaching technologies that presume to help create interaction, social presence, and community in online educational venues. Discussion regarding the theoretical underpinnings and factors that influence the development and implementation of these techniques is sparse. Published literature regarding online teaching strategies often focuses on more conventional technologies such as e-mail and computer-mediated conferencing (Moisey, Neu, & Cleveland-Innes, 2008). Other literature centres on emerging technologies on a macro level, such as virtual worlds, social networking technologies, learning management systems, and presence pedagogy (Bronack, Cheney, Riedl, & Tashner, 2008).

This chapter explores a specific group of emerging educational technologies that our preliminary research has shown may help enhance interaction, facilitate a shared social experience, and create a culture of community in online post-secondary classes. We call these "artistic pedagogical technologies" (APTs) and define them as teaching strategies founded in the arts. Typically they may include literary, visual, musical, or dramatic elements. APTs are distinguished from traditional online technologies in part by their emphasis on aesthetics and their link to creativity. This discussion of APTs takes the conversation regarding emerging technologies for online education to a micro level as we examine specific teaching strategies.

Background

Exemplary online educators infuse a sense of presence in the online classes they teach (Perry & Edwards, 2005). This sense of presence is both created and conveyed through the incorporation of interactive teaching strategies such as Photovoice, virtual reflective centres, and conceptual quilting (Perry, 2006; Perry, Dalton, & Edwards, 2008). Preliminary research found that these teaching technologies helped the students directly and helped to stimulate interaction between students and teachers, between students, and between students and the course materials. The result of such interactions is the enhancement of the experience of social presence in the virtual class, creating what we have labelled a "culture of community" (Perry & Edwards, 2009, 2010). The repeated experience of an authentic shared presence helps to establish shared values, norms, and beliefs; a shared culture in the online class. This chapter builds on this research and examines a category of emerging teaching technologies; APTs. Previously unpublished findings regarding APTs are featured, APTs are further defined and described, and we speculate on how and why these approaches may make online teaching more effective. This discussion is framed from the theoretical standpoint of Vygotsky's (1978) Social Development Theory (SDT). Theoretical and pedagogical shifts that could influence the development, adoption, or use of APTs are described.

Current Relevant Literature

A plethora of literature supports the importance of interaction, social presence, and community in online education (Conrad, 2002; Garrison, 2007; Lee, Carter-Wells, Glaeser, Ivers, & Street, 2006; Liu, Magjuka, Bonk, & Lee, 2007; Rovai & Wighting, 2005; Shu-Fang & Aust, 2008). The literature is less forthcoming in terms of how to facilitate APTs.

Interaction

Moore defines interaction in online education as a student-course content, student-student, or student-teacher exchange (1989). Others add interaction between student and self (Ornelles, 2007; Sorensen, Takle, & Moser, 2006) and between student and technology (Battalio, 2007). Anderson (chapter 2) further expands the notion of interaction including individuals, technology, and content. Positive outcomes of interaction in online courses include creativity and collaboration (Zwirn, 2005), increasing higher-order thinking and retention (Bevis, 1989), and moving online courses away from being text-based correspondence classes (Shea, Pickett, & Pelz, 2003).

Social presence

Social presence is the ability of students and teachers to project their personal characteristics into the online class, thereby presenting themselves as "real people" (Rourke et al., 2000). The value of social presence to effective online teaching and learning is commonly highlighted. For example, social presence is one cornerstone of the widely supported Community of Inquiry Model (Rourke et al., 2000).

Kehrwald (2008) cautions that despite the general agreement among researchers that social presence is a key element in effective online teaching and learning, a shared understanding of social presence has not yet emerged. However, there does seem to be general agreement that interaction in the online classroom is linked to the experience of social presence (Kehrwald, 2008; Perry et al., 2008; Perry & Edwards, 2010; Rourke et al., 2000). Shea et al. (2003) concluded that successful teachers who engage fully with learners from a distance use teaching strategies that stimulate interaction by conveying human presence.

Community

The effective online classroom is a social community (Swan, 2003) that enacts community values such as the exchange of beliefs and ideas (Marzano, 1998). We define community as shared culture in the online classroom, including shared values, norms, and beliefs (Perry & Edwards, 2009, 2010). Others have defined community as a classroom in which knowledge is mutually constructed (Abbott & Fouts, 2003; Peterson, Carpenter, & Fennema, 1989).

The creation of an online learning community serves as the foundation for a successful learning environment (Conrad, 2002; Lee et al., 2006). Learners in such a community are active and engaged (Bandura, 2000; Rice-Lively, 1994), experience enhanced self-worth (Conrad, 2002) and increased cognitive learning (Bandura, 2000; Liu et al., 2007; Rovai, 2002b), do not experience alienation and isolation (Knowles, 1990; Moule, 2006; Rice-Lively, 1994; Rovai, 2002b; Rovai & Wighting, 2005; Saritas, 2008), and finish programs and courses (Wiesenberg & Stacey, 2005, 395). Moule found "mutual engagement," "joint enterprise," and "shared repertoire" resulted from what she called an "e-community" (pp. 133–139). Moisey et al. (2008) found significant positive correlations between students' satisfaction with their courses and programs and levels of the sense of community cohesion. Rovai (2002a) argued that online classrooms have the same potential to build and sustain a sense of community as do face-to-face classes. He stated that a learning community "consists of four related dimensions: spirit, trust, interaction, and commonality of learning expectations and goals" (p. 12).

Facilitation of interaction, social presence, and community

Facilitating interaction, social presence, and community in the online classroom is primarily the teacher's responsibility (Swan, 2003). Rovai agrees, saying, "given the particular affective nature of forming and maintaining a sense of community in the online classroom, extra demands are placed on … facilitators" (2002b, p. 3).

There is minimal literature related to specific teaching strategies that facilitate these goals. In an analysis of three credible distance education journals seeking to identify trends in research related to interaction in Internet-based distance learning, Karatas (2008) reported that between 2003 and March 2005, there were no articles published on design-related topics such as instructional strategy development and course materials design. Educators are left to create interactive teaching technologies to achieve these goals, yet the literature suggests that they are often not successful (Pelz, 2004). Pelz noted that most online courses are still isolating, one-way, correspondence courses (2004). Often teaching strategies are developed and utilized without being first subjected to rigorous research-based assessment (Perry & Edwards, 2005).

Some researchers provide broad guidelines for pedagogy that could enhance interaction, social presence, and community in online courses. For example, activities that promote negotiation and debate (Ouzts, 2006), teacher communication behaviours that reduce social and psychological distance (Shu-Fang & Aust, 2008), mimicking proximity by addressing social and psychological factors such as social space and social presence (Hodge et al., 2006), and dialogue that allows knowledge to be constructed, deconstructed, and reconstructed (Bakhtin, 1986; Wegerif, 2006) are proposed. Rovai suggests that "instructors teaching at a distance may promote a sense of community by attending to seven factors: transactional distance, social presence, social equality, small group activities, group facilitation, teaching style and learning stage, and community size" (2002a, p. 12).

Studies make reference to the importance of immediacy (communication behaviours that reduce social and psychological distance between people) in facilitating interaction, social presence, and community online (Kreijns, Kirschner, & Jochems, 2002; Na Ubon & Kimble, 2004; Richardson & Swan, 2003; Woods & Baker, 2004). These investigations focus on conventional tools such as computer conferencing systems (CCSs), online chats, or e-mail (Moisey et al., 2008; Saritas, 2008). Saritas found that CCSs enhanced social interaction, collaboration, and dialogue, and Moisey et al. found that CCSs had positive effects on online community cohesion (2008).

In summary, interaction, social presence, and community are widely accepted as important to effective online teaching and learning. Educators are often without evidence-based guidance as to what teaching technologies will help to facilitate these goals. Artistic pedagogical technologies seem to help accomplish these outcomes in online post-secondary classrooms (Perry & Edwards, 2005; Perry et al., 2008). How and why this happens is not yet fully understood.

Definition and Description of Artistic Pedagogical Technologies

Artistic pedagogical technologies

Online instructors need to develop, implement, and evaluate new and creative teaching technologies to maximize interaction, social presence, and community in the virtual class. Our team published preliminary findings related to three such teaching technologies (Photovoice, virtual reflective centres, and conceptual quilting) demonstrating positive educational outcomes (Perry & Edwards, 2005; Perry & Edwards, 2006; Perry et al., 2008). Specifically, both students and teachers reported that their virtual classrooms were effective learning environments, in part because of the inclusion of these teaching technologies (Perry, 2006; Perry et al., 2008). Students reported benefitting scholastically from the sense of community that arose when they participated in these learning activities. One finding from our preliminary studies that requires further analysis is the link between Photovoice, virtual reflective centres, and conceptual quilting teaching strategies — they are all founded in the arts (visual arts and drama). Why do artistic approaches, which value aesthetics as well as reason (Gull, 2005), seem to facilitate community in the online class?

The worth of the arts has been recognized in face-to-face education. Specifically, art, photography, literature, poetry, music, and drama have been reported as contributing positively to the face-to-face classroom educational experience by stimulating reflection and helping to achieve affective objectives (Brett-McLean, 2007; Brown, Kirkpatrick, Magnum, & Avery, 2008; Calman, 2005; Darbyshire, 1994; Gull, 2005; Mareno, 2006; Reilly, Ring, & Duke, 2005; Wright, 2006). However, only one of these reports (Darbyshire, 1994) is research based. Darbyshire found that face-to-face arts-based teaching strategies create a safe environment that stimulates dialogue.

The translation of artistic-based pedagogy to the online classroom seems to be an untested idea. Brown, Kirkpatrick, Magnum, and Avery (2008) declare there is a need to move from established online pedagogies that no longer fully satisfy today's learner and to "develop and implement alternative interpretative pedagogies" (p. 283). Skiba (2006) concludes that "This generation views learning as a social and constructive activity that must be experiential, engaging, interactive, and collaborative" (p. 103). These qualities seem well matched to APTs.

Photovoice

Our research team studied Photovoice in several research pilot projects (Perry, 2006; Perry & Edwards, 2005; Perry et al., 2008). Wang and Burris (1997) developed Photovoice as a participatory-action research methodology. Perry and Edwards transformed this research methodology into an interactive online teaching technology, which involves the instructor posting a digital image and a reflective question at the onset of each unit in the course. Students are encouraged to discuss the question in a dedicated forum. Photovoice is non-graded and optional.

Positive outcomes included encouraging engagement and interest in the course content; making the learning environment more appealing, creative, and interesting; and facilitating the development of social cohesiveness (Perry et al., 2008).

Virtual reflective centres

An example of an APT that involves the artistic element of drama is the virtual reflective centre (Ronaldson, 2004). Virtual reflective centres are role-playing simulation exercises that are reported to enhance critical thinking and promote social presence online (Ronaldson, 2004). Cubbon (2008) performed trial virtual reflective centres in an online graduate course for advanced nursing practice students. Cubbon randomly assigned students to either a patient or a nurse practitioner role and gave each student information needed to fulfill the roles during a real-time online "appointment." As a summation, the instructor distributed reflective questions related to the exercise and hosted an asynchronous group discussion.

Participants in the virtual reflective centre exercise emphasized that it facilitated the development of a sense of community in this virtual classroom because it provided a safe, structured environment in which they could engage in an interactive learning exercise. Students commented that the dramatic element of the exercise helped to make the activity novel and engaging, which motivated socially meaningful interaction (Cubbon, 2008).

Conceptual quilting

Conceptual quilting was developed by the authors and has been used in online graduate courses as a summary activity. Students are asked to construct a virtual quilt that is comprised of ideas, metaphors, theories, and other details from the course that they found most meaningful. The "quilt" needs to be in a medium that can be shared electronically with the class.

The construction of the conceptual quilt encourages learners to reflect as they interact again with course materials. Further interaction with the instructor and other students comes when students post their quilts to an asynchronous online discussion forum and respond to comments. This often results in a resurgence of dialogue around a course theme that was depicted in the quilt. The activity is non-graded and optional. However, participation is almost 100 percent. Anecdotally, students comment that conceptual quilting helps them consolidate their learning and bring closure to the course. From a social interactive perspective, the sharing of the completed quilts is a way for students to acknowledge the impact that others (teachers and peers) have had on their learning.

How Artistic Pedagogical Technologies Encourage Interaction, Create Social Presence, and Facilitate a Sense of Community

We propose that the educational impact of these arts-based teaching technologies arises initially because of the enhanced interactions they help create. The interpersonal interactions among students and between students and teachers, and the intrapersonal interactions between students and self are most relevant to this discussion. These interactions may lead to the experience of social presence, as those in the virtual classroom reveal elements of their personal characteristics and become more "real" and known to one another and to themselves. Social presence cannot be established, indeed cannot exist, without interpersonal and intrapersonal interactions. These do not necessarily take place spontaneously in virtual classrooms. Specific teaching technologies that have social interaction (leading to social presence) as a goal are needed to facilitate this outcome.

Not all social presence is equal. Some social presence is more authentic, perhaps experienced as more "human" or "real" by participants. The quality of social presence that is generated through APTs is described by students as palpably "human." Because APTs are founded in the arts, which are very human- centred (created by, valued by, shared by, and appreciated by people), they help to facilitate interpersonal and intrapersonal social presence that is less artificial.

Not all interactions are alike in terms of effect on social presence and the eventual formation of community. Frequency of interaction alone is not an adequate assessment of interaction levels. While the number of times that students interact with peers, teachers, course materials, and themselves may be important, it is the quality of those interactions that may be most critical to positive outcomes such as social presence and community. For example, a brief e-mail exchange containing superficial greetings exposes little of the values, attitudes, or beliefs of participants. To be meaningful to the establishment of social presence and community, interactions must reveal something important and relevant about participants to others or to oneself.

Further, social presence in the online class needs to be part of a course from the beginning to the end. That is, participants need to establish their initial presence when the course begins, but they also need to demonstrate ongoing participation in the course (Kehrwald, 2008). Teaching technologies such as Photovoice that require student and teacher contributions throughout the course may help facilitate both becoming known to each other at the beginning of a course and provide ongoing evidence of participation. Further, APTs such as Photovoice potentially allow participants to systematically reveal more of their personal values, beliefs, and priorities as the course proceeds. This may facilitate progressively more personal and perhaps more authentic and meaningful social interaction.

Essentially effective social presence in the online class is a dynamic experience. It evolves over the duration of the course with participants becoming more comfortable with one another through ongoing meaningful interactive experiences. Eventually this leads to the establishment of a culture of community.

Kehrwarld concluded that the establishment and growth of social presence is related to three conditions: ability, opportunity, and motivation (2008). APTs help to meet each of these conditions. First, *ability* refers to students being able to reference their own experiences and bring these to the learning community in an appropriate way. Kehrwarld emphasizes that novice learners do not come to online classes with this skill; they may not have the ability to send and to read social presence cues. Students need learning activities that help them to gain this ability. For example, with Photovoice, students are given a specific non-threatening invitation to share something of themselves. Photovoice becomes both the vehicle for students to establish their social presence in the course, and — because the same strategy is used often in the course — a strategy that teaches students how to share socially in the online milieu. Participants also model this skill for one another, and those students who may be unskilled at sending and reading social presence cues have the option of waiting, watching, and learning how to participate prior to contributing.

The second condition is opportunity for interaction. Opportunities need to be purposefully created in online courses to facilitate frequent meaningful interactions helping to cultivate social presence. Because APTs are used on a regular basis in a course (in the case of Photovoice Weekly), there is a consistent, scheduled opportunity for participants to interact. While opportunities for interaction are easy to create, they need to be such that learners are not overwhelmed by the demands of interaction within large groups (Harrison & Thomas, 2009; Heejung, Sunghee, & Keol, 2009). Most APTs, such as virtual reflective centres, are suited to smaller class sizes to allow for participation by all students. The Photovoice activity requires students to make one or two short responses. Long responses with references are discouraged in this activity. This keeps participants from being overwhelmed by a large number of long posts they feel obliged to respond to.

Teaching technologies that require students and teachers to contribute in a visible way signal that they are available for interactions (Kehrwald, 2008). APTs all have a tangible element that provides these signals. In the case of Photovoice, the evidence of the participation of the teacher is the weekly posting of the photo. Evidence of student involvement is the responses to the Photovoice question. Likewise, the conceptual quilts posted by students are evidence that they are members of a specific educational community. The responses and questions raised in reaction to the quilts are evidence of "attendance" and the involvement of other class community members.

The third condition for the establishment and growth of social presence is motivation. Teaching tools need to motivate students to participate. Motivation often comes because students believe that participation has some benefit for them. If the activity creates interest, motivation may be enhanced. For example, the Photovoice activity has mysterious elements (one student commented that she never could guess what photo would be hidden under the "electronic paperclip"), arousing curiosity and motivating participation. We speculate that perhaps part of what makes Photovoice motivational is that students find it engaging. It catches their attention; one student described it as a "hook" that captured her interest. Once students are focused on the course theme, the Photovoice activity engages them in dialogue with themselves as they puzzle over the image and think about their response to it. Because there is no correct response to art, their reaction — out of necessity — must be personal. Then, as the class members begin to share their personal responses to the image in the public forum, there is some social expectation (motivation) to reciprocate by doing the same, and a public dialogue results in meaningful social interaction.

Students may be demotivated if they believe excessive time and effort is required to participate. There is no requirement to participate in Photovoice or conceptual quilting, which allows students to lurk without participating. Without exception, in our experience, over the time of the courses all students eventually regarded the Photovoice exercise as worthwhile, and contributed. Keeping class sizes reasonable helps to prevent participants from being overwhelmed by the number of postings related to each Photovoice activity. Students receive positive feedback from peers and instructors regarding their participation in these activities, fuelling motivation.

Vygotsky's (1978) Social Development Theory (SDT) helps explain how APTs influence interaction, social presence, and the creation of a culture of community in the online class. Teaching and learning, whether occurring in a traditional or virtual classroom, are essentially social experiences. According to SDT, social interaction is fundamental to cognitive development. Consciousness and cognition result from socialization and social behaviour. Vygotsky focused on the connections between people and the socio-cultural context in which they act, and interact, in shared experiences (Hung, 2001). SDT learning is characterized by mediation through language, the discovery of differing perspectives, and the achievement of shared meaning (2001). Vygotsky's SDT promotes learning environments in which students play an active role in learning. Teachers, rather than being transmitters of knowledge, collaborate with students to facilitate the acquisition of new knowledge, skills, and attitudes. Learning becomes a reciprocal experience involving oneself and others.

When educators apply SDT to online education, learners require effective teaching tools to facilitate interacting from a distance, particularly with teachers and other students. When effective teaching strategies are used, online learners can achieve social connections with other students and teachers that, according to SDT, facilitate learning (Perry & Edwards, 2006).

We propose that APTs stimulate these authentic human interactions required to promote social engagement in the virtual class. For example, music, artistic images, and literary works are infused with the humanness of the composer, artist, or author. When APTs are part of, or the foundation for, a course activity, they introduce into the course some aspect of another human. While a traditional learning activity in an online course may appear rather barren and anonymous, a song, photograph, or poem is often infused with the values, preferences, and beliefs of the one who created it. We suggest that when another "real" person is introduced into the online course using an APT, the potential for human interaction is enhanced. From the students' perspective, now there is someone to interact with.

The stimulation provided by the inclusion of such a strategy seems to be a catalyst for interaction for several reasons. One respondent in a study involving the use of Photovoice wrote, "Seeing a new photographic image appear each week in my course forum was like seeing the artwork that might be displayed in my professor's home. It told me something about her, about how she saw the world. It made her more real somehow and made it comfortable for me to e-mail her and ask questions." Another student respondent offered a comment that helps to further the explanation regarding how the inclusion of an APT in a course stimulated meaningful interaction, saying, "I felt like I got to know my professor because of the type of photos that were included in the course. I could tell that she had an appreciation for nature … and probably had a kind heart. I participated more freely because I felt like I knew her from the photos."

To achieve genuine, appropriate, authentic, interaction that results in substantive discussion, debate, and reflection may require deliberate strategies on the part of the online teacher. We propose that the inclusion of APTs in online course design may precipitate engagement between students, and students and teachers, which — according to SDT — is necessary for learning.

APTs provide an opportunity for meaningful interpersonal and intrapersonal interaction. APTs require a contribution that provides class members evidence of the involvement of students and teachers in a course. Ongoing meaningful interactions facilitate authentic social presence, which lays the foundation for and facilitates the ongoing development of the culture of community. In a culture of community, participants embrace shared values, norms, and beliefs; a shared culture. A shared culture facilitates further meaningful interpersonal interactions, and the cycle is propelled (Figure 7.1).

Figure 7.1 Development of a culture of community in the online classroom

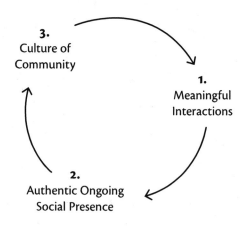

Factors Influencing the Development, Adoption, and Use of Artistic Pedagogical Technologies

Shea identified three foundational changes that have influenced online education: a philosophical shift from objectivism towards constructivism; a theoretical shift from behaviorism towards socio-cognitive views of education; and a pedagogical shift from direct instruction to the facilitation of collaborative learning (2006). Shea argued that these changes encourage teaching approaches that help to develop virtual learning communities (2006). For example, student-centred, learner-directed, interactive, participative pedagogical methods are congruent with the establishment of community in the online class, with social interaction, and ultimately with learning. It follows that the development, adoption, and use of online teaching strategies, in this case APTs, is influenced by these factors.

From objectivism to constructivism

Objectivists emphasize the accumulation of facts, and view learners as passive recipients of knowledge (Kelly, 1970). Differing views and individual experiences are often discouraged (Gulati, 2008).

Constructivists embrace different worldviews and emphasize social relationships and cognitive interaction in learning environments (Goodyear, 2002; Hung & Chen, 2001; Larochelle & Bednarz, 1998). Teaching technologies that encourage learners to construct knowledge through activity and experience are favoured (Jonassen, 1999) over lectures.

Online learning environments are excellent venues for constructivist teaching technologies (Kehrwald, 2008). The potential for connectivity afforded by online communications facilitates opportunities for human-human interaction that, according to constructivists, precipitates learning. APTs such as Photovoice, conceptual quilting, and virtual reflective centres all purposefully create social interaction. In keeping with a constructivist philosophy, such interactive learning may involve the modification of attitudes, beliefs, and knowledge in all participants, including students and teachers. Such modification has been described as transactional (Shin, 2002) or interactivist (Bickhard, 1992).

From behaviourism to socio-cognitivism

Behaviourism focuses on observable and measurable behaviours (Good & Brophy, 1990). For example, Bloom's (1956) taxonomy of learning is the basis for the development of behavioural learning objectives in which learning tasks are broken down into specific measurable tasks. For behaviourists, the achievement of objectives equates with learning success.

Cognitive theorists view learning as involving internal processes, such as comparing new information to existing knowledge. This makes learning more active and complex. Learning strategies such as metaphors, chunking information, and the organization of instructional materials from simple to complex are used by cognitivists to facilitate learning.

APTs would be viewed favourably by cognitivists. Photovoice activities require students to engage in higher-order thinking, asking that they compare something they know to the theory of the course. For example, if the image presented is a photo of a tree with leaves changing colour, and the topic in the course is factors that influence organizational change, students are asked to recall what they know about weather, light, temperature, and seasonal influences on trees in the autumn, and to translate this into determining factors within an organization that might also create change. An internal thought process is needed, as changes in nature become a metaphor for changes in organizations. Likewise, in conceptual quilting students use internal mental processes to seek and find relationships between key themes in the course, and to find ways to weave these together in meaningful patterns that they can then display and explain.

From direct instruction to collaborative learning

The hallmark of direct instruction is teacher control with one-way transmission of information and measurable learning. Collaborative learning involves joint intellectual efforts by students or students and teachers as they work together to seek understanding, meaning, or solutions. Students depend on and are accountable to one another as they participate in learning activities, and there is usually an end product to the learning activity.

Both a prerequisite for and a result of collaborative learning online may be the establishment of a community of learners. According to Ascough, "one of the key features of an online course is the employment of activities that will allow students to get to know one another better" (2002, p. 13).

APTs can facilitate collaborative learning. For example, virtual reflective centres involve the active participation of all students, as each is assigned a role and invited to participate in a shared experience. Participants depend on one another to play their parts so the activity succeeds. Similarly, in a Photovoice activity, while students initially contribute their own interpretations of the photo, the resulting online discussion becomes a collaborative learning activity as learners work together to formulate common understandings of the relationships between the photo and course topics.

APTs are congruent with the emerging constructivist philosophy of learning. As online educators come to appreciate more diverse ways of knowing and understanding, as we focus more on social relationships in the class, and as we shift from a "world of facts to a world of symbols and models" (Larochelle & Bednarz, 1998, p. 7), educational technologies that have a human element, such as APTs, may become popular.

Conclusion

This chapter provided a new understanding regarding emerging online teaching strategies, specifically, artistic pedagogical technologies. Teaching strategies founded in the arts may assist online educators who aim to make online courses more meaningfully interactive. With meaningful interaction comes the potential for the experience of authentic ongoing social presence and the eventual establishment of a culture of community.

At present, there is limited development of such teaching tools, and research on those that have been developed is in its infancy (see chapter 1). The explanations presented in this chapter regarding why APTs are effective teaching strategies are also only a beginning point. The potential educational impact of such teaching technologies (on students and teachers) has not yet been wholly explored. This chapter contributes to these discussions and encourages educators, course designers, and researchers to experiment with including aspects of the arts in learning activities in online courses.

REFERENCES

Abbott, M., & Fouts, J. (2003). *Constructivist Teaching and Student Achievement: The Results of a School-level Classroom Observation Study in Washington.* Lynnwood, WA: Seattle Pacific University.

Ascough, R. (2002). Designing for online distance education: Putting pedagogy before technology. *Teaching Theology and Religion, 5*(1), 17–30.

Bakhtin, M. (1986). *Speech Genres and Other Essays.* Austin, TX: University of Texas Press.

Bandura, A. (2000). Exercise of human agency through collective efficacy. *Current Directions in Psychological Science, 9,* 75–78.

Battalio, J. (2007). Interaction online: A reevaluation. *Quarterly Review of Distance Education, 8*(4), 339–352.

Bevis, E.O. (1989). Teaching and learning: A practical commentary. In E.O. Bevis & J. Watson (Eds.). *Toward a Caring Curriculum: A New Pedagogy for Nursing.* New York, NY: National League for Nursing.

Bickhard, M. (1992). How does the environment affect the person? In L. Wineger & J. Valsiner (Eds.). *Children's Development in Social Context* (pp. 33–52). Mahwah, NJ: Lawrence Erlbaum.

Bloom, B. (1956). *Taxonomy of Educational Objectives.* NY: David McKay.

Brett-McLean, P. (2007). Use of the arts in medical and health professional education. *University of Alberta Health Sciences Journal, 4*(1), 26–29.

Brown, S.T., Kirkpatrick, M.K., Magnum, D., & Avery, J. (2008). A review of narrative strategies to transform traditional nursing education. *Journal of Nursing Education, 47*(6), 283–286.

Bronack, S.C., Cheney, A.L., Riedl, R.E., & Tashner, J.H. (2008). Designing virtual worlds to facilitate meaningful communication: Issues, considerations, and lessons learned. *Technical Communication, 55*(3), 261–269.

Calman, K.C. (2005). The arts and humanities in health and medicine. *Public Health, 119,* 958–9.

Conrad, D. (2002). Deep in the hearts of learners: Insights into the nature of online community. *Journal of Distance Education, 17*(1). Retrieved 10 March 2010, from http://cade.athabascau.ca/vol117.1/conrad.html

Cubbon, J. (Under Review). Use of online role play to enhance motivational interviewing skills of Advanced Nursing Practice graduate students. *Journal of Nursing Education.*

Dabbagh, N. (2004). Distance Learning: Emerging pedagogical issues and learning designs. *The Quarterly Review of Distance Education, 5*(1), 37–49.

Darbyshire, P. (1994). Understanding caring through arts and humanities: A medical/nursing humanities approach to promoting alternative experiences of thinking and learning. *Journal of Advanced Nursing, 19*, 856–863.

Garrison, D.R. (2007). Online community of inquiry review: Social, cognitive, and teaching presence issues. *Journal of Asynchronous Learning Networks 11*(1) 61–72. University of Calgary. Retrieved 18 November 2008, from http://www.communitiesofinquiry.com/documents/Community%20of%20Inquiry%20Issues.pdf

Good, T., & Brophy, J. (1990). *Educational Psychology: A Realistic Approach.* (4th ed.). White Plains, NY: Longman.

Goodyear, P. (2002). Psychological foundations for networked learning. In C. Steeples & C. Jones (Eds.), *Networked Learning: Perspectives and Issues* (pp. 49–76). London: Springer.

Gulati, S. (2008). Compulsory participation in online discussions: Is this constructivism or normalisation of learning? *Innovations in Education and Teaching International, 45*(2), 183–192.

Gull, S.E. (2005). Embedding the humanities into medical education. *Medical Education, 39*, 235–6.

Harrison, R., & Thomas, M. (2009). Identity in online communities: Social networking sites and language learning. *International Journal of Emerging Technologies and Society, 7*(2), 109–124.

Heejung, A., Sunghee, S., & Keol, L. (2009). The effects of different instructor facilitation approaches on students' interactions during asynchronous online discussion. *Computers and Education, 53*(3), 749–760.

Hodge, E., Bosse, M.J., Faulconer, J., & Fewell, M. (2006). Mimicking proximity: The role of distance education in forming communities of learning. *International Journal of Instructional Technology and Distance Learning.* Retrieved November 19, 2007, from http://www.itdl.org/Journal/Dec_06/article01.htm

Hung, D. (2001). Theories of learning and computer-mediated instructional technologies. *Education Media International, 38*(4), 281–87.

Hung, D., & Chen, D. (2001). Situated cognition, Vygotskian thought and learning from the communities of practice perspective: Implications for the design of web-based e-learning. *Education Media International, 38*(1), 3–12.

Jonassen, D. (1999). Designing constructivist learning environments. In C. Reigeluth (Ed.), *Instructional Theories and Models* (pp. 215–240). Mahwah, NJ: Lawrence Erlbaum.

Karatas, S. (2008). Interaction in the Internet-based distance learning researches: Results of a trend analysis. *The Turkish Online Journal of Educational Technology, 7*(2).

Kehrwald, B. (2008). Understanding social presence in text-based online learning environments. *Distance Education, 29*(1), 89–106.

Kelly, G. (1970). A brief introduction to personal construct theory. In D. Bannister (Ed.), *Perspectives in Personal Construct Theory* (pp. 1–29). London: Academic Press.

Knowles, M. (1990). *The Adult Learner: A Neglected Species.* Houston, TX: Gulf.

Kreijns, K., Kirschner, P., & Jochems, W. (2002). The sociability of computer-supported collaborative learning environments. *Educational Technology and Society, 5*(1), 8–22.

Larochelle, M., & Bednarz, N. (1998). Constructivism and education: Beyond epistemological correctness. In M. Larochelle, N. Bednarz, & J. Garrison (Eds.), *Constructivism and Education* (pp. 3–20). Cambridge, UK: Cambridge University Press.

Lee, J., Carter-Wells, J., Glaeser, B., Ivers, K., & Street, C. (2006). Facilitating the development of a learning community in an online graduate program. *The Quarterly Review of Distance Education, 7*(1), 13–33.

Mareno, N.A. (2006). A nursing course with the great masters. *Nursing Education Perspectives, 27*(4), 182–3.

Marzano, R. (1998). *A Theory-based Meta-analysis of Research on Instruction.* Aurora, CO: Midcontinent Regional Educational Laboratory.

Moisey, S., Neu, C., & Cleveland-Innes, M. (2008). Community building and computer-mediated conferencing. *Journal of Distance Education, 22*(2), 15–42.

Moore, M. (1989). Three types of interaction. *The American Journal of Distance Education, 3*(2), 1–6.

Moule, P. (2006). Developing the communities of practice, framework for on-line learning. *Electronic Journal of eLearning, 4*(3), 133–140.

Na Ubon, A., & Kimble, C. (2004). Exploring social presence in asynchronous text-based online learning communities (OLCS). In *Proceedings of the 5th International Conference on Information Communication Technologies in Education 2004* (pp. 292–297), Samos Island, Greece. Retrieved 22 July 2008, from http://www-users.cs.york.ac.uk/~kimble/research/icicte.pdf

Ornelles, C. (2007). Providing classroom-based intervention to at-risk students to support their academic engagement and interactions with peers. *Preventing School Failure, 51*(4), 3–12.

Ouzts, K. (2006). Sense of community in online courses. *The Quarterly Review of Distance Education, 7*(3), 285–296. Retrieved 2 November 2007, from http://o-web.ebscohost.com.aupac.lib.athabascau.ca/ehost/pdf?vid=3&hid=104&sid=b8254076-4ede-4f7c-9949-46a8720b2147%40sessionmgr107

Pelz, B. (2004). Three principles of effective online pedagogy. *Journal of Asynchronous Learning Networks, 8*(3), 33–46.

Perry, B. (2006). Using photographic images as an interactive online teaching strategy. *The Internet and Higher Education, 9*(3), 229–240.

Perry, B., & Edwards, M. (2005). Exemplary online educators: Creating a community of inquiry. *Turkish Online Journal of Distance Education, 6*(2), 46–54.

Perry, B., & Edwards, M. (2006). Exemplary educators: Creating a community of inquiry online. Invited publication in the *International Experience in Open, Distance, and Flexible Education Collection of Papers from the Open and Distance Learning Association of Australia (ODLAA) Conference,* November 2005.

Perry, B., & Edwards, M. (2009). Strategies for creating virtual learning communities. *Encyclopaedia of Information Science and Technology* (2nd ed.). Hershey, PA: IGI Global.

Perry, B., & Edwards, M. (2010). Interactive teaching technologies that facilitate the development of online learning communities in nursing and health studies graduate courses. *Teacher Education Quarterly, 37*(1).

Perry, B., Dalton, J., & Edwards, M. (2008). Photographic images as an interactive online teaching technology: Creating online communities. *International Journal of Teaching and Learning in Higher Education, 20*(2), 106–115.

Peterson, P., Carpenter, T., & Fennema, E. (1989). Teachers' knowledge of students' knowledge in mathematics problem-solving: Correlational and case analyses. *Journal of Educational Psychology, 81,* 558–569.

Reilly, J.M., Ring J., & Duke, L. (2005). Visual thinking strategies: A new role for art in medical education. *Family Medicine, 37*(4), 250–2.

Rice-Lively, M. (1994). Wired warp and woof: An ethnographic study of a networking class. *Internet Research, 4*(4), 20–35.

Richardson, J., & Swan, K. (2003). Examining social presence in online courses in relation to students' perceived learning and satisfaction. *Journal of Asynchronous Learning Networks, 7*(1), 68–88.

Ronaldson, S. (2004). Untangling critical thinking in educational cyberspace [Unpublished doctoral dissertation]. University of British Columbia, Vancouver, BC, Canada.

Rourke, G., Garrison, D., Anderson, T., & Archer, W. (2000). Critical inquiry in a text-based environment: Computer conferencing in higher education. *The Internet and Higher Education, 2*(2–3), 87–105.

Rovai, A.P. (2002a). Building sense of community at a distance. *International Review of Research in Open and Distance Learning, 3*(1). Retrieved 16 May 2008, from http://www.irrodl.org/index.php/irrodl/article/viewFile/79/153

Rovai, A.P. (2002b). Sense of community, perceived cognitive learning, and persistence in asynchronous learning networks. *The Internet and Higher Education, 5,* 319–332.

Rovai, A.P., & Wighting, M.J. (2005). Feelings of alienation and community among higher education students in a virtual classroom. *The Internet and Higher Education, 8*, 97–110.

Saritas, T. (2008). The construction of knowledge through social interaction via computer-mediated communication. *Quarterly Review of Distance Education, 8*(1), 35–49.

Shea, P. (2006). In online environments. *Journal of Asynchronous Learning Network, 10*(1). Retrieved 22 August 2008, from http://www.sloan-c.org/publications/jaln/v10n1/v10n1_4shea.asp

Shea, P., Pickett, A., & Pelz, W. (2003). A follow-up investigation of "teaching presence" in the SUNY learning network. *Journal of Asynchronous Learning Networks, 7*(2), 61–80. Retrieved 7 November 2007, from http://www.aln.org/publications/jaln/v7n2/pdf/v7n2_shea.pdf

Shin, N. (2002). Beyond interaction: The relational construct of "transactional presence." *Open Learning, 17*(2), 121–137.

Shu-Fang, N., & Aust, R. (2008). Examining teacher verbal immediacy and sense of classroom community in online classes. *International Journal on E-Learning, 7*(3), 477–498.

Skiba, D.J. (2006). Collaborative tools for the net generation. *Nursing Education Perspectives, 27*(3), 162–3.

Sorensen, E., Takle, E., & Moser, H. (2006). Knowledge-building quality in online communities of practice: Focusing on learning dialogue. *Studies in Continuing Education, 28*(3), 241–257.

Swan, K. (2003). Examining social presence in online courses in relation to students' perceived learning and satisfaction. *Journal of Asynchronous Learning, 7*(1).

Vygotsky, L.S. (1978). *Mind and Society: The Development of Higher Mental Processes.* Cambridge, MA: Harvard University.

Wang, C., & Burris, M. (1997). Photovoice: Concept, methodology, and use for participatory needs assessment. *Health Education Behaviour, 24*, 369–387.

Wegerif, R. (2006). Dialogic education: What is it and why do we need it? *Education Review, 19*(2), 58–66.

Wiesenberg, F., & Stacey, E. (2005). Reflections on teaching and learning online: Quality program design, delivery, and support issues from a cross- global perspective. *Distance Education, 26*(3), 385–404.

Woods, R., & Baker, J. (2004). Interaction and immediacy in online learning. *International Review of Research in Open and Distance Learning.* Retrieved 2 November 2007, from http://www.irrodl.org/content/v5.2/woods-baker.html

Wright, D.J. (2006). The art of nursing expressed in poetry. *Journal of Nursing Education, 45*(11), 458–461.

Zwirn, E.E. (2005). Using media, multimedia, and technology-rich learning environments. In D.M. Billings & J.A. Halstead (Eds.), *Teaching in Nursing: A Guide for Faculty* (pp. 377–396). St. Louis, MO: Elsevier.

STRUCTURED DIALOGUE EMBEDDED WITHIN EMERGING TECHNOLOGIES

8

> *Yiannis Laouris, Gayle Underwood, Romina Laouri,*
> *& Aleco Christakis*

Acknowledgements

This work would not have been possible without the enthusiastic contributions of Structured Dialogic Design pioneers around the globe and especially those collaborating with the Institute for 21st Century Agoras: Kevin Dye, Ken Bausch, Roy Smith, Tom Flanagan, Peter Jones, Norma Romm, Janet Mcintyre and Larry Fergeson. The examples cited were made possible through a United Nations Development Program (UNDP; "Building a multi-ethnic and multi-national Cyprus to promote European values and regional and international peace") and a European initiative ("Uniting for citizenship and participation: Youth promoting vulnerable groups' rights, opportunities, and knowledge") given to the Future Worlds Center. The authors would also like to thank the coordinators of the projects mentioned here; Mr. Larry Fergeson and Mrs. Kerstin Wittig; the participants of the co-laboratories for all their constructive suggestions; and also CWA Ltd. for providing their proprietary software Cogniscope2™ for use in all co-laboratories.

Abstract

The science of Structured Dialogic Design (SDD) is embedded within emerging technologies to develop a new, scientifically grounded methodology in online distance education. The chapter begins with an introduction of the SDD process. It then discusses current applications of wikis in educational contexts and their shortcomings. Examples in which the SDD was embedded within emerging technologies and wikis in particular are used to draw attention to the benefits introduced by the application of SDD as a tool to structure the learning process and facilitate commitment, endurance, and intentionality of learning.

Introduction

Contrary to common belief, most learning does not take place in formal settings. Schooling is an invention of the last centuries, whereas learning existed long before evolution presented Homo sapiens (Laouris & Eteokleous, 2005). Learning takes place at all times because we carry the necessary equipment in our skull; we can compare on-going stimuli, experiences, and judgments with existing schemata and assimilate them as new knowledge (Duncan, 1995; Laouris, 1998a, 2004a, 2005; Laouris & Eteokleous, 2005). Moreover, learning takes place without our conscious realization because the brain can process, evaluate, and organize information asynchronously at the time of the input. Information technology and the Internet, in connection with the exponential proliferation of mobile telecommunications technologies, have taken the learning process outside of school walls, providing access to knowledge to people from all walks of life (Laouris & Laouri, 2008). Such technologies have created new venues for learning to take place anytime, anywhere. Technology improves connectivity among learners, including between management, parents, and students, as well as future generations. Furthermore, new and emerging technologies impact pedagogy and teaching and learning processes (see chapters 1, 2, 5, 6). For example, emerging technologies have irreversibly shifted the balance from teacher-centred towards learner-centred education (chapter 5). Educational systems and pedagogical theories are continuously evolving (chapter 2) and are increasingly integrating state-of-the-art technologies, transforming "distance education." Immersed within technology-rich environments, learners and educators may communicate at all times synchronously and asynchronously. This imposes new requirements for learning theories, pedagogy, and andragogy (chapter 2; Knowles, 1984). Prior to educators having time to adapt to technological changes, the appearance of Web 2.0 tools has created strong turbulence. Wikis, blogs, podcasts, talking characters, and virtual environments with 3-D avatars "living" almost normal lives present new challenges to educators. The plethora of tools, in conjunction with conjectures and disputes concerning their effectiveness and educational relevance, introduce a new "state of affairs": the emergence of a mosaic of independent approaches and technologies to access an unstructured and almost chaotic body of content and knowledge (chapters 2, 5).

It is within this context that our team focuses its efforts in the development and application of methodologies that can adequately address the above state of affairs. Over the last five years, our global team has collaborated to facilitate a marriage between the openness and freedom of Web 2.0 tools and the structure and discipline that education requires (Christakis & Underwood, 2008; Laouris & Christakis, 2007). In the next sections we (a) introduce the reader to the science of Structured Dialogic Design, which serves as the scientific grounding of a new approach to education that has recently been implemented in distance education contexts, (b) review contemporary approaches that educators use to integrate wikis in their educational settings, and (c) discuss example applications of the Structured Dialogic Design process embedded into a particular Web 2.0 tool, specifically the wiki.

A Short Introduction to the Science of Structured Dialogic Design

The Structured Dialogic Design (SDD) process is a self-documenting and strictly structured method of disciplined and democratic dialogue between people. The SDD is scientifically grounded on seven laws of complex systems science and cybernetics (see Laouris, Laouri, & Christakis, 2008, p. 340). A typical SDD co-laboratory is specifically designed to assist heterogeneous groups of individuals to deal with complex issues in a reasonably limited amount of time (Banathy, 1996; Christakis, 1996; Warfield, 1994; Warfield & Cardenas, 1994). It enables the integration of contributions from learners or stakeholders with divergent prior knowledge and with diverse backgrounds and perspectives. This integration is achieved through a process that is structured, inclusive, and collaborative.

The need for a scientific methodology to facilitate democratic dialogue was first envisioned by systems thinkers in the Club of Rome (Özbekhan, 1969, 1970). SDD was systematically refined through years to its current 4th-generation version, which has a much wider applicability. (The interested reader can refer to the complete review of the methodology by Christakis and Bausch (2006) for more details.) Laouris & Christakis (2007) reviewed the first four applications of the 4th-generation SDD (referred to as "hybrid") that were implemented in the context of a rich web-based communication environment using a combination of asynchronous and synchronous communication tools. The term "hybrid" is used to describe the fact that the SDD process is implemented using (a) a combination of face-to-face and virtual communication technologies, and (b) a combination of synchronous (which can be either face-to-face or virtual) and asynchronous sessions. Subsequently, SDD scientists have extended the hybrid model to integrate wiki technologies to support the first phases of the process: the collection, clarification, and discussion of learners' contributions implemented through a wiki. The wiki offers a self-documenting mechanism and also serves as a shared space where deliberations can continue after the completion of an SDD process. (For more information on the processes for a typical hybrid virtual and face-to-face SDD co-laboratory, see Laouris et al., 2008; Laouris et al., 2007; Laouris & Michaelides, 2007; and Laouris, Michaelides & Sapio, 2007.)

What Are Wikis and How Are They Currently Used?

The purpose of a wiki is to function as a Web page that is quick to edit and can be used as a shared space of collaboration. A wiki can be set up so that a user can easily add, remove, edit, and change the content of the page. One of the most important features of a wiki is its "history" feature, which allows wiki owners to view all previous edits to the wiki, along with the edits of respective authors. If needed, the wiki can be "rolled back" to a previous state. This ease of interaction and operation makes wikis effective tools for mass collaborative authoring. Augar, Raitman, and Zhou (2004, 2005, 2006) note that wikis have two different styles of usage. The first is known as *document mode.* In document mode, learners create collaborative documents (chapter 11). Multiple authors can edit and update the content of a document. Gradually the content becomes a representation of the shared knowledge or beliefs of the contributors (Leuf & Cunningham, 2001). The second wiki style is known as *thread mode.* Contributors carry out discussions in the wiki environment by posting signed messages. Others respond, leaving the original messages intact. Eventually a group of threaded messages evolves (chapter 14). Wikis can also have two states, *read* and *edit.* Wikis are in a "read" state by default (i.e., a wiki page looks like a Web page). When a user wishes to edit a page, s/he must access the wiki's "edit" state.

Since their inception, wikis have found application in education as Computer Supported Collaborative Learning (CSCL) tools. For example, Leuf & Cunningham (2001) describe wikis being used at Georgia Tech University to facilitate CSCL. The Georgia Tech wiki, known as CoWeb, enabled students to create documents as a group, review articles and post comments, create informational resources similar to Wikipedia, and disseminate information among the student body. Augar et al. (2004) describe work at Deakin University with wiki applications such as hosting an icebreaker exercise to facilitate ongoing interaction between members of online learning groups. Further examples of wiki uses are presented in chapters 11 and 14.

The combination of ease of use and potential for collaboration is making wikis a powerful distance education tool. Current applications in educational spaces include: student collaboration, exploring new projects, and opening the "classroom."

Student collaboration

Wikis represent a place for learners to work in small groups (Underwood, 2008). The shy, quiet student is "heard" on the wiki through his/her contributions. The student who is always the first to contribute will not receive undue attention or become frustrated for never being chosen, as his/her contributions are combined with the rest of the group on the wiki. Instructors can easily monitor the progress of group work, and post helpful hints and reminders. The result may be a compendium of information about a topic that students can access whenever they need to do so. Teachers and students may also post their course notes on wikis. Some students may do so to help their peers. Others do so because they were asked to do so by their instructor. Some may feel that this is an instance of cheating, while others may feel that this is an example of open learning. The true educational value occurs when student additions extend and clarify the information given by the instructor. The advantage to this kind of collaboration is the fact that the collectively authored notes represent an amount of knowledge that is more than the sum of the knowledge that each individual had before the process (referred to as an "emergent property" in the science of complex systems). For example, students may come to know what their peers know best, their different perspectives, and whom to "consult" on different issues. Furthermore, students have the opportunity to ask each other questions to clarify meanings and concepts.

Exploring new projects

This application is probably most suited for the wiki whenever there is a topic that is new to the class and to the instructor. Students and instructor explore the topic together and *collectively create a knowledge base* on the wiki. They "teach" one another and "interact" on an equal playing field, and together co-create an understanding of a topic. Such activities require an innovative instructor who is willing to step out of his/her role as "keeper of the knowledge" and step into the role of learner along with the students, while still supervising, monitoring and imposing structure (chapter 14).

Opening the "classroom"

Wikis provide opportunities for openness in the classroom. Possibilities arise to have other learners from the same institution or from another country "visit" and review and critique projects. For example, Couros (chapter 6) discusses opening his classroom to his personal learning network. Vicki Davis (from the U.S.) and Julie Lindsay (from Qatar) present the Flat Classroom Project (n.d.), which uses a wiki to join two classrooms into one large virtual classroom where middle- and high-school students from two different countries collaborate on projects throughout the school year. The Palestinian-German Twinning School Programme (n.d) offers similar virtual communications to enable children in Gaza to interact with German peers. In 1998 in Cyprus, Hrach Gregorian and the first author of this chapter collaborated with the International Communication and Negotiation Simulation Project (ICONS) of the University of Maryland to allow Turkish- and Greek-Cypriots to participate in virtual negotiation workshops at a time when crossing the border was not possible (Kaufman, 1998; Laouris, 2004b).

Wikis Embedded within a Structured Dialogic Design Process

The purpose of the following sections is to describe recently implemented applications, which demonstrate the combined application of wikis and/or other synchronous/asynchronous communication technologies embedded within the SDD process. The goals on which we focus our efforts include:

> exploit the power of wikis and offer a smooth transition between current wiki uses and the integration of SDD;
> design a *hybrid* learning environment in the sense of enabling synchronous and asynchronous interactions between learners;
> enable *structured* and *focused* collaboration between learners;
> ensure that the *process remains structured and controlled*; and
> ensure that the *process can be completed* and learners will reach a well-defined goal.

Application of SDD to facilitate educational reforms

The SDD method has been used in settings that aim to facilitate educational reforms. For example, in 2007, the State of Michigan, decided to include universal design for learning (UDL) in their educational curriculum. UDL is a framework for designing educational environments that enable diverse populations of learners to gain knowledge, skills, and enthusiasm for learning. To help support the systemic change needed for UDL, the State of Michigan established a referent group of diverse stakeholders, including many from general education, to develop a shared vision for Michigan with regard to meeting the needs of diverse learners (Christakis, Coston, & Conway, 2007). During the referent group's dialogic deliberations in three two-day participative events, called co-laboratories, the participants identified (a) idealized requirements, (b) barriers, and (c) opportunities for developing learning community models complementary and compatible with the principles of UDL. The SDD process was employed to enable the diverse group of stakeholders to use democratic planning to address complex, boundary-spanning challenges at the state level.

One year later, a small school in Michigan used a streamlined virtual Internet version of the SDD called Webscope© that enabled multiple stakeholders at the local school district to participate in a mixed-presence approach (Bausch et al., 2008; Underwood & Christakis, 2008). The stakeholders were busy educators from a school district who did not have the time or money to meet face to face for 6 days. The mixed-presence disciplined dialogue took place using a Webscope© wiki approach and enabled participants to discover the root causes of the issue of high dropout rates for their school district. The stakeholders were also able to recognize the complexity and multi-dimensionality of the problem confronting their community.

All the participants reported that they had had no experience using a wiki prior to this project. They were familiar with using e-mail to communicate with each other, and half of them had participated in a Web seminar in the past. The one-hour Web seminar on how to use the Webscope© wiki appeared to be sufficient to get them up to speed on how to use a wiki. The only suggestions/corrections given while participants collaborated online were related to the disciplined process as opposed to the actual use of the wiki technology. For example, during Round 2 of the Webscope©, participants were reminded to refrain from making judgments or comments but, instead, to just ask clarifying questions.

When using the Webscope© wiki, participants typed comments and questions into online dialogues as they learned from each other about their individual perceptions of the factors contributing to the dropout rate. Although a schedule was followed, the flexibility of the wiki allowed everyone to contribute while they were in their own place at their own time. For instance, participants reported that some of their wiki entries were completed while they sat in their living room, with their children playing nearby; other entries occurred while team members were at a conference or on vacation. In essence, the Webscope© wiki was a virtual democratic space (see chapter 14) where participants met to discuss the high dropout rate issue.

While meeting face to face for disciplined dialogue is most effective, all participants agreed that, given their busy schedules, engaging with the first four rounds online saved time and was a convenient way for their group of stakeholders to actively participate. Additionally, the virtual SDD significantly improved the efficiency of producing results by offering stakeholders the opportunity to interact at different times from different places. Preliminary estimates indicate that virtual SSDs will reduce the cost of face-to-face meetings by a factor of six by minimizing participants' travel and per diem expenses (Christakis & Underwood, 2008).

Analogous co-laboratories have been organized in Cyprus with United Nations and European Union funding. For example, the United Nations Development Program funded an initiative running under the title "Building a multi-ethnic and multi-national Cyprus to promote European values and regional and international peace." This project used structured dialogue in five elementary schools in Cyprus to assist participants in developing a vision for a multicultural transformation of their schools (Multicultural schools, n.d.). The process engaged not only teachers and parents, but also young pupils. It was probably the first time that Cypriot youth (12 years old) participated on an equal footing in designing the schools of the future. Because of practical difficulties and busy schedules, the SDD process was implemented as a hybrid of synchronous and asynchronous sessions. The results have shown that the Structured Dialogic Design was instrumental in empowering and liberating young participants to contribute significant ideas. Figure 8.1 shows the vision map constructed by the participants of one elementary school, with about half the ideas coming from youth. It is worth noting how the collective wisdom of a small number of pupils, teachers, and

Figure 8.1 Vision map constructed by pupils, teachers, and parents of one Cypriot elementary school

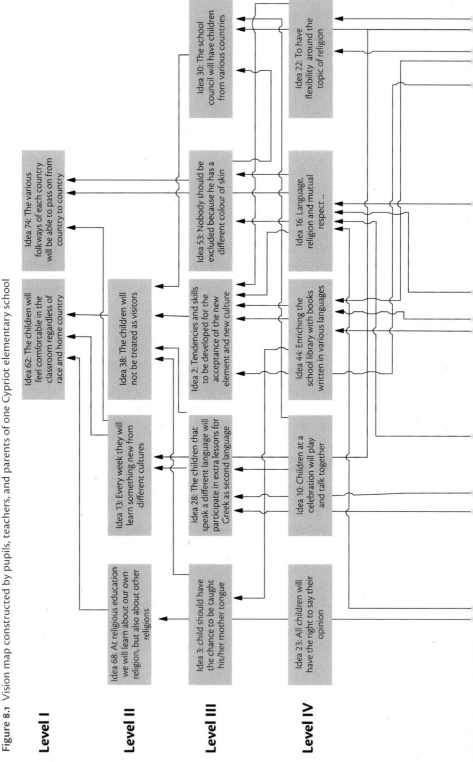

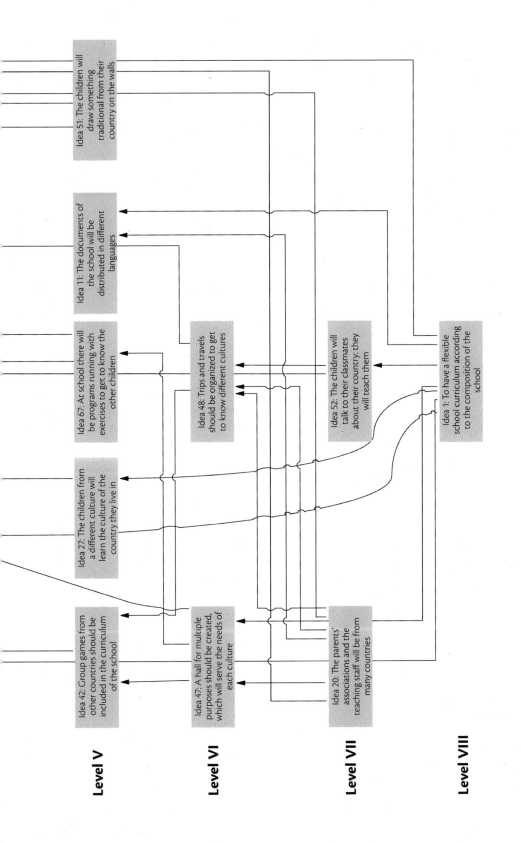

Idea 51: The children will draw something traditional from their country on the walls

Idea 11: The documents of the school will be distributed in different languages

Idea 67: At school there will be programs running with exercises to get to know the other children

Idea 48: Trips and travels should be organized to get to know different cultures

Idea 52: The children will talk to their classmates about their country: they will teach them

Idea 1: To have a flexible school curriculum according to the composition of the school

Idea 27: The children from a different culture will learn the culture of the country they live in

Idea 42: Group games from other countries should be included in the curriculum of the school

Idea 47: A hall for multiple purposes should be created, which will serve the needs of each culture

Idea 20: The parents' associations and the teaching staff will be from many countries

Level V

Level VI

Level VII

Level VIII

parents resulted in a detailed map and important ideas surrounding educational reforms. At the end of the SDD, the participants decided to continue with the envisioning and implementation of practical actions designed towards materializing their vision.

Application of SDD to support group learning among youth representatives

The next example demonstrates the application of the SDD process to support group learning. The participants were twenty representatives of youth organizations from eleven European countries. Their goal was to identify the reasons why youth across Europe do not actively participate in European projects and affairs. Youth were engaged in three consecutive SDDs, which took place in Cyprus, Italy, and Romania. Their ideas were partly collected by e-mail and partly documented in wikis. The resulting tables with all ideas were categorized in clusters, and the resulting map of their first co-laboratory can be found online at the project's wiki (http://ucyvrok.wetpaint.com). Through group work, participants seemed to significantly expand their knowledge and understanding of the situation: whereas one individual could come up with three to seven ideas, all participants together came up with an average of one hundred ideas (in each co-laboratory). Additionally, the structured process facilitated a gradual evolution in their thinking such that the youth did not only adopt ideas coming from others, but critically evaluated and enhanced their own ideas as well. Through the process of exploring the influences of one idea on another, they agreed on which ideas might be more influential than others when designing problem solutions. This project is ongoing, and the wiki serves as a continuing collaborative space. The final results will be forwarded to the European Parliament, the body that first raised this concern. The SDD process has made it possible to collect ideas from twenty individuals across eleven countries to be used by a legal body such as the EU Parliament so that targeted actions can be taken.

Discussion

Educators around the world show increasing interest on the concept of "collaborative" or "peer" learning. The benefits of peer learning have long been recognized (Jonassen, 1994), and recent research about peer learning focuses mainly on collaboration, communities of practice, mentorship, and other models of peer interaction (Hansman, 2008). Yet despite the increased interest and expectations to use peer learning in classrooms, we are still missing appropriate methods and tools. Much of the published work that examines the effects of peer learning stems from the social psychological construct of interdependence among members of a group (O'Donnell, 2002). Strategic and effective use of peer learning in the classroom requires teachers to understand the implications of a variety of theoretical approaches so that they can adapt their use of peer learning to the demands of the task. The application of Structured Dialogic Design is a promising tool towards this direction.

Wikis play a central role in the majority of applications that embrace collaborative learning. Raitman, Augar, and Zhou (2005) revealed the following as the greatest disadvantages of the application of wikis in education: (1) its faceless contact was not personal enough for real interactions to take place; (2) lack of discussion; (3) user-interface lacked simplicity; (4) pages are too long to scroll; (5) lack of real-time communication; (6) too easy to delete someone else's contributions. Some of these disadvantages have been resolved because wiki technology since 2005 has advanced to meet user needs. However, the structured dialogue embedded within a hybrid Webscope© wiki satisfactorily addresses the challenges posed above: the fact that asynchronous interactions are complemented with synchronous plenary sessions renders the contact between individual learners more personal. The SDD process ensures structured communication: For example, the different phases of the SDD process allow learners to (a) engage in questions and clarifications through the wiki (i.e., Clarification Phase), (b) present and support their points of view, (c) engage in direct discussions, and (d) select and vote across contributions. Additionally, specific features of the SDD process lead towards valued goals. For instance, at times during the process, learners are prohibited from making value statements and criticizing the ideas of others. This facilitates the creation of an environment of mutual respect and trust (Tsivacou, 1997).

Current trends in education shift more towards experiential, problem-based, just-in-time types of learning. For example, Knowles (1975) postulated long ago that (adult) learning must be problem-centred, rather than content-oriented. As a result of a nation-wide experiment, the authors also proposed that learning should focus on content that has immediate relevance to learners' personal lives (Laouris, 1998b; Laouris & Anastasiou, 2005). The fact that the SDD process gives participants the freedom to contribute their ideas satisfies this requirement.

Reaching a shared understanding under the SDD process

The ultimate goal of a learning community is to reach a state in which all learners achieve a deep and shared understanding of the problem at hand (Law of *Requisite Meaning*, Turrisi, 1997). During the first rounds of the SDD (Figure 8.2), participants share their previous knowledge and their different points of view. Peer learners are encouraged to request clarifications, but they are not allowed to make any value statements regarding the statements or contributions of others. The authenticity and autonomy of each participant is "protected" through compliance with SDD rules that call for respect and tolerance (Law of *Requisite Autonomy in Decision*, Tsivacou, 1997). Upon completion of the initial phases, learners are expected to expand their explicit knowledge of the issue at hand due to information sharing and collaboration. The different types of knowledge acquired and refined during these phases are then further elaborated. Although ideas are formed in the minds of individuals, interaction between individuals within an SDD process typically plays a critical role in developing these ideas, in line with Vygotsky's (1978) theory of constructivism. In other words, "communities of interaction" contribute to the amplification and development of new knowledge. At later stages of the SDD process, learners explore and compare ideas, exploring new viewpoints and discovering new perspectives. The nautilus spiral (Figure 8.3) offers a good visualization model borrowed from nature, which corresponds to the phases of a typical SDD process: during the first circle of an SDD, the number of ideas increases quickly (the separators in the nautilus spiral lie close to one another). All participants expand their explicit knowledge about the issue. In subsequent circles, ideas develop in "depth" and quality and not in numbers. Learners achieve a much deeper understanding, as illustrated in the nautilus spiral, with the nautilus seperators increasing in surface.

Through adherence to the laws of Structured Dialogic Design, we cultivate autonomy, facilitate evolutionary learning (Dye & Conway, 1999), and assist participants in achieving meaning and wisdom. Out of these largely cognitive processes, action emerges as a natural consequence (Law of *Requisite* Action; Laouris, Laouri, & Christakis, 2008), which translates to commitment, endurance, and intentionality of learning. The SDD process not only facilitates better learning, it contributes to rendering learners self-driven and more enthusiastic, therefore serving the learner-centred principle.

Figure 8.2 The spiral of learning. Knowledge is acquired in incremental phases. During each subsequent phase of the SDD process, learners acquire meaning and wisdom in an evolutionary manner.

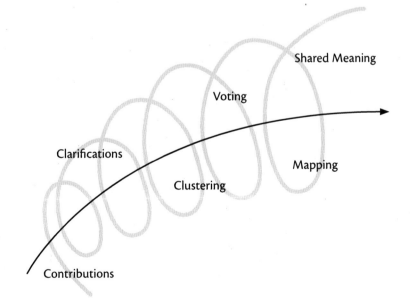

Figure 8.3 Nautilus spiral (This Wikipedia and Wikimedia Commons image is from the user Chris 73 and is freely available at http://commons.wikimedia.org/wiki/File: NautilusCutawayLogarithmicSpiral.jpg under the creative commons cc-by-sa 2.5 license.)

Obstacles and inhibitors to the application of the SDD process in learning environments

The greatest obstacle to the application of the SDD in a learning context arises when learners are not expected to cooperate. Worse, when individualistic learning is valued more than collaborative learning, SDD is not the method of choice. An interdependent group is one in which the learning outcomes of its members are linked: in a truly cooperative interdependent group, no one can succeed unless everyone succeeds (Johnson, Johnson, & Holubec, 1992). The SDD process is, by design, such a group learning method. Learners learn more when they learn collectively (Johnson & Johnson, 2005). Indeed, the SDD process is inappropriate for individualistic learning (even though Bausch [2000] applied SDD to collect, condense and prioritize the principles or standards that govern the practice and ethics of design). An inhibitor closely related to the above lies in current assessment practices. If assessment gives learners the message that only individual achievement is valued, and that collaborative work is akin to cheating, then the potential of collaborative learning will never be realized. If learners are predisposed to this way of thinking, they might refrain from contributing during the process, thus seriously restraining the outcome.

In an SDD setting, the emphasis is in the application of *reciprocal peer learning*, in which students act both as teachers and learners. While there is recognition in the literature that peer learning can contribute to the social and psychological needs of learners (Griffiths, Houston, & Lazenbatt, 1995; Slavin, 1995), many authors tend to treat peer learning mainly as an instructional strategy, rather than an approach which pursues a broader educational agenda, as is so prominent in SDD settings (e.g., Multicultural schools, n.d).

The SDD process is also inhibited when one or more experts view themselves as possessing knowledge that needs to be transferred to learners. The SDD process is based on the assumption that a significant amount of knowledge already exists in the minds of the learners. However, no one learner possesses all required knowledge. Additionally, individual differences exist: learners' expertise and backgrounds may be different. The SDD process enables teachers and learners to efficiently contribute to each other's knowledge base using a structured approach, and to harness collective wisdom to co-produce meaning and wisdom without inhibiting or limiting individual learning.

Finally, the application of the SDD process might be hindered in situations where learners are not comfortable with the process, required technologies, and virtual environments. In a recent experiment, we engaged an international group of SDD experts to work together towards discovering the roadblocks facing President Barack Obama in realizing his vision of a bottom-up democracy (www.obamavision.wikispaces. com). This project highlighted some of the inhibitors and symptoms of technological/computing literacy.

Conclusions and Future Developments

Distance education virtual environments offer incredible opportunities for educators to engage learners in a variety of learning experiences. At the same time, "unstructured" environments pose new challenges (see chapter 14). Furthermore, some technologies (e.g., wikis) have an asynchronous character, while others (e.g., Multi-user Virtual Environments such as the one described in the chapter 15) are synchronous. Without the classic teacher-student roles (see chapter 3), it is becoming increasingly difficult to provide basic instruction and guidance. Distance educators need to explore new ways of teaching (chapter 2) by capitalizing upon the multi-faceted nature of new media, rather than by simply translating existing face-to-face techniques into the new media. Theories of education cannot simply be transferred in new learning environments. Since emerging technologies for education transcend academic disciplines (chapter 1), it is also necessary that we develop new theories of education and learning that account for diverse constraints and challenges. The SDD approach offers a theoretical grounding that is promising. SDD has recently been used in the context of distance education for learners to explore solutions to societal issues and concerns. However, the method has not yet been tested for diverse types of content and learning: our work has focused on complex societal problems while traditional content areas such as physics, biology, economics, and mathematics have not yet been investigated. The SDD approach is particularly useful for problems that are complex and for which learners might have different perspectives and possibly conflicting interpretations. However, as discussed above, the method also faces shortcomings. In the future, it is desirable that the method is tested in diverse educational settings.

REFERENCES

Augar, N., Raitman, R., & Zhou, W. (2004). Teaching and learning online with wikis. In R. Atkinson, C. McBeath, D. Jonas-Dwyer, & R. Philips (Eds.), *Beyond the Comfort Zone: Proceedings of the 21st ASCILITE Conference* (pp. 95–104). Perth. Retrieved from http://www.ascilite.org.au/conferences/pertho4/procs/augar.html

Augar, N., Raitman, R., & Zhou, W. (2005). Wikis: collaborative virtual learning environments. In J. Weiss, J. Nolan, J. Hunsinger, & P. Trifonas (Eds.), *The International Handbook of Virtual Learning Environments* (pp. 1251–1269). Netherlands: Springer.

Augar, N., Raitman, R., & Zhou, W. (2006). Developing wikis to foster web-based learning communities: An iterative approach. *International Journal of Web Based Communities, 2* (3), 302–317.

Banathy, B. (1996). *Designing Social Systems in a Changing World*. New York, NY: Plenum.

Bausch, C.K. (2000). The practice and ethics of design. *Systems Research and Behavioral Science , 17*, 23–50.

Bausch, C.K., Underwood, G., & Christakis, A.N. (2008). *Addressing the Dropout Rate in Two Michigan School Districts. Report generated by G. Underwood for the Participants at the Dropout Rate for Evart on June 5–26.* Evart, MI: 21st Century Agoras.

Christakis, A.N. (1996). A people science: The Cogniscope system approach. *Systems Journal of Transdisciplinary Systems Sciences, 1*(1), 18–25.

Christakis, A.N., & Bausch, K.B. (2006). *How People Harness Their Collective Wisdom and Power to Construct the Future.* Greenwich, CT: Information Age Publishers.

Christakis, A.N., & Underwood, G. (2008). Virtual co-laboratories for implementing effective teaching strategies in local schools. *International Cybernetic Society Conference.* Barcelona, Spain: International Cybernetic Society.

Christakis, A.N., Coston, C., & Conway, D. (2007). *Developing a Learning Community for Universal Design for Learning in Michigan Schools.* Michigan: 21st Century Agoras.

Duncan, R.M. (1995). Piaget and Vygotsky revisited: Dialogue or assimilation? *Developmental Review, 15,* 458–472.

Dye, K.M., & Conway, D.S. (1999). *Lessons Learned from Five Years of Application of the CogniScope Approach to the Food and Drug Administration.* Paoli, Pennsylvania: CWA Ltd. Interactive Management Consultants.

eTwinning project. (n.d.). Retrieved 30 December 2008, from http://www.etwinning.net/en/pub/index.htm

Fergesson, L., & Laouris, Y. (n.d.). NML Wiki Report. Retrieved from http://cyprusmedia.wetpaint.com/page/SDDP+Report

Flat Classroom Project. (n.d.). Retrieved 30 December 2008, from http://flatclassroomproject.Wikispaces.com

Griffiths, S., Houston, K., & Lazenbatt, A. (1995). *Enhancing Student Learning Through Peer Tutoring in Higher Education.* Coleraine, Ulster: Educational Development Unit, University of Ulster.

Hansman, C. (2008). Adult learning in communities of practice: Situating theory in practice. In C. Kimble, P. Hildreth, & I. Bourdon (Eds.), *Communities of Practice: Creating Learning Environments for Educators.* Charlotte, NC: Information Age Publishing.

Johnson, D.W., & Johnson, R.T. (2005). New developments in social interdependence theory. *Genetic, Social, and General Psychology Monographs, 131*(4), 285–358.

Johnson, D.W., Johnson, R.T., & Holubec, E. (1992). *Advanced Cooperative Learning.* Edina, MN: Interaction Book Company.

Jonassen, D. (1994, April). Thinking technology: Towards a constructivist design model. *Educational Technology,* 34–37.

Kaufman, J.P. (1998). Using Simulation as a Tool to Teach About International Negotiation. *International Negotiation*, 3:1, 59–75.

Knowles, M. (1975). *Self-Directed Learning.* Chicago: Follet.

Knowles, M. (1984). *Andragogy in Action.* San Francisco: Jossey-Bass.

Laouris, Y. (1998a). Computer-brain spirit: Where are their borders. In P. Savides & E. Savides (Eds.), *You Tell Me, Third Research Symposium of Art & Design* (pp. 39–42). Nicosia, Cyprus: Art Studio Laboratories.

Laouris, Y. (1998b). Innovative education for the new millennium. In M. Nikias (Ed.), *A Leap into the New Millennium.* Nicosia, Cyprus: IMSC.

Laouris, Y. (2004a). Meta-thoughts about the code used by the brain. *Brain, Mind, and Culture: Neuroscience Forum* (p. 50). Limassol, Cyprus: RIKEN Brain Science Institute.

Laouris, Y. (2004b). Information technology in the service of peace building: The case of Cyprus. *World Futures*, 60(1 & 2), 67–79.

Laouris, Y. (2005). The computer has elicited a new evolutionary step in the development of the human mind. *(draft available on request).* Nicosia, Cyprus: Future Worlds Center.

Laouris, Y., & Anastasiou, H. (2005). The introduction of IT in the lives of children as a service to global peace: Experiences from a nation-wide experiment 15 years after. *Proceedings of the 4th World Conference on Mobile Learning, mLearn.* Cape Town. Retrieved from www.mlearn.org.za/CD/papers/Laouris%20&%20Anastasiou.pdf

Laouris, Y., & Christakis, A. (2007). Harnessing collective wisdom at a fraction of the time using Structured Design Process embedded within a virtual communication context. *International Journal of Applied Systemic Studies*, 1(2), 131–153.

Laouris, Y., & Eteokleous, N. (2005). An educationally relevant and socially responsible definition of mobile learning. *Proceedings of the 4th World Conference on Mobile Learning, mLearn.* Cape Town. Retrieved from http://www.mlearn.org.za/CD/papers/Laouris%20&%20Eteokleous.pdf

Laouris, Y., & Laouri, R. (2008). Can information and mobile technologies serve close the gap and accelerate development? *World Futures*, 64(4), 254–275.

Laouris, Y., & Michaelides, M. (2007). What obstacles prevent practical broadband applications from being produced and exploited? In P. Roe (Ed.), *Towards an Inclusive Future Impact and Wider Potential of Information and Communication Technologies* (pp. 281–299). Brussels: COST.

Laouris, Y., Laouri, R., & Christakis, A.N. (2008). Communication praxis for ethical accountability: the ethics of the tree of action: dialogue and breaking down the wall in Cyprus. *Systems Research and Behavioral Science*, 25(2), 331–348.

Laouris, Y., Michaelides, M., & Sapio, B. (2007). What are the obstacles that prevent the wide public from benefiting and participating in the broadband society? *Proceedings of the Cost 298 Conference.* Moscow.

Leuf, B., & Cunningham, W. (2001). *The Wiki Way: Quick Collaboration on the Web.* Upper Saddle River, NJ: Addison Wesley.

Multicultural schools. (n.d.). Retrieved 30 December 2008, from www.multiculturalcyprus.net

O'Donnell, A.M. (2002). Promoting thinking through peer learning. *Theory Into Practice, 41*(1), 2–4. Retrieved 30 December 2008, from http://findarticles.com/p/articles/mi_moNQM/is_/ai_90190482?tag=artBody;col1

Özbekhan, H. (1969). Towards a general theory of planning. In E. Jantsch (Ed.), *Perspectives of Planning.* Paris: OECD Publications.

Özbekhan, H. (1970). On some of the fundamental problems in planning. *Technological Forecasting, 1*(3), 235–240.

Palestinian-German School Twinning Programme. (n.d.). Retrieved 30 December 2008, from http://www.school-tp-ps.de

Raitman, R., Augar, N., & Zhou, W. (2005). Employing wikis for online collaboration in the e-learning environment: Case study. *Proceedings of the 3rd International Conference on Information Technology and Applications.*

Slavin, R.E. (1995). *Cooperative Learning.* Boston: Allyn and Bacon.

Tsivacou, I. (1997). The rationality of distinctions and the emergence of power: a critical systems perspective of power in organizations. *Systems Research and Behavioral Science, 14*(1), 21–34.

Turrisi, P.A. (1997). *Pragmatism as a Principle and Method of Right Thinking.* New York, NY: State University of New York Press.

Underwood, G. (2008). *What is a wiki? Collaboration on the Web.* Retrieved 30 December 2008, from http://mits07.pbWiki.com/f/What%20is%20a%20Wiki.pdf

Underwood, G., & Christakis, A.N. (2008). *Diagnosis of the root causes for the drop out rate of Evart district schools.* Retrieved 30 December 2008, from http://effectiveteaching.wikispaces.com/file/view/Final+report+summary+for+Evart+June29+2008.pdf

Vygotsky, L.S. (1978). *Mind and Society: The Development of Higher Mental Processes.* Cambridge, MA: Harvard University Press.

Warfield, J.N. (1994). *A Science of Generic Design.* Ames: Iowa University Press.

Warfield, N. & Cardenas, A.R. (2004). *A Handbook of Interactive Management.* Ames, IA: Iowa State University Press.

Part 3
Social, Organizational,
and Contextual Factors
in Emerging Technologies
Implementations

9 PERSONAL LEARNING ENVIRONMENTS

> *Trey Martindale & Michael Dowdy*

Abstract

The concept of the personal learning environment has emerged in recent years via the work of online theorists, researchers, and developers. This emergence is the result of (1) the limitations experienced by administrators, trainers, teachers, and learners using learning management systems (LMSs), and (2) the recognition of the importance of informal, lifelong, and "lifewide" learning. A PLE has been conceptualized as both a broad, holistic learning landscape and as a specific collection of tools that facilitate learning. In this chapter we will discuss the brief history of the PLE, why the PLE is useful, PLE examples, the PLE compared with the LMS, objections and barriers to the PLE, and directions for the future of the PLE.

Introduction

Seekers of knowledge in today's world have plenty of options beyond institutional courses or formal classroom-based training sessions. The World Wide Web is a resource that creates the potential for profound learning experiences compared to those achieved through traditional courses and classrooms. Many methods for improving learning have been explored in the past decade, and a common thread is the use of new technologies to facilitate learning. Constructivist learning models describe the value of learners making meaning of their own experiences (Wilson & Lowry, 2000). Web-based resources have the potential to enable constructivist learning environments. However, when learners have access to a practically limitless repository of information, it can be challenging to create meaning from that information. The challenge is not to provide access to information but to provide a framework for making sense of the information.

As the Web has evolved as an information resource and medium, Web tools and processes have also evolved. The term "Web 2.0" has been used to describe this evolution of the Web from an information source to a "read/write" medium (O'Reilly, 2005). The development of Web 2.0 technologies has given learners a large collection of tools, sometimes called social software, for creating, organizing, and making meaning from content (chapters 2, 3, 4, 14). Social software has a long history, and can be defined simply as software that supports group interaction (Allen, 2004). Web users can now interact with Web content as well as with other users in a shared environment that was not possible just a few years ago. Using such software, learners can organize content that has meaning to them and easily share that content and their own interpretation of it. Further, learners can interact with other people with shared learning goals. This new interplay among learners and between learners and content has not reached the status of a consensual definition or understanding. However, the concept of the personal learning environment (PLE) is one way to describe this type of Web-facilitated learning environment.

The PLE certainly qualifies as an emerging technology as defined in the opening chapter of this volume (chapter 1). The PLE is a somewhat new and evolving construct, has gone through at least one hype cycle, is not yet fully understood, and is potentially disruptive with unfulfilled potential. The concept of the PLE has been emerging in recent years via the work of online theorists, researchers, and developers, as the result of the limitations of learning management systems (LMSs), a recognition of the importance of informal and lifelong learning, and the growth of social software. In this chapter we will discuss the brief history of the PLE, why the PLE is useful, PLE examples, a comparison of the PLE to the LMS, objections and barriers to the PLE, and directions for future PLE development.

PLE Defined

The PLE concept has emerged from discussions among a wide-ranging group of professionals interested in designing and supporting online learning environments. At present, no single environment or application instantiates an archetypal PLE. For some, a PLE is a specific tool or defined tool collection used by a learner to organize his or her own learning processes. For others, the PLE simply acts as a metaphor to describe the activities and milieu of a modern online learner. Much like other concepts within this volume (e.g., see chapters 1, 3, and 4), there is not a widely accepted definition of the PLE. However, one common trait in all the early definitions of a PLE is that the PLE gives the learner control over his or her own learning process. Because the PLE idea has developed in part as a reaction to learning management systems, it is not surprising to see "personal" control represented in descriptions of a PLE.

The phrase "personal learning environment" appears to have first been mentioned at the annual JISC-CETIS (Joint Information Systems Committee — Centre for Educational Technology Interoperability Standards) conference in 2004 (Schaffert & Hilzensauer, 2008). The development history of the PLE concept has been documented in resources such as Wikipedia (History of personal learning environments, 2008) and by Mark van Harmelen of the University of Manitoba (van Harmelen, 2008). We refer readers to these two sources for more detail on the history of PLEs. A key event in PLE history was Scott Wilson's presentation of "the VLE of the future" (Wilson, 2005). Soon afterward, the PLE was a theme of the 2005 JISC-CETIS annual conference.

As the PLE idea gained exposure, researcher Scott Leslie solicited and posted a collection of PLE models (Leslie, 2008) that would receive a great deal of attention. Ray Sims included an interesting PLE diagram (see http://simslearningconnections.com/ple/ray_ple.html) that highlighted not only Web 2.0 technologies but also personal relationships. Sims included meditation, books read, and the physical spaces where he learns (office, bicycling in his local area, the library, and home). This highly personalized version adds a dimension to PLEs beyond social networking technologies.

Educational technologist David Warlick's PLE diagram incorporated "reflective endeavours" that included reading, writing, giving presentations, and conversing with practitioners. The reflective endeavours were not oriented towards or dependent upon specific technologies to facilitate interaction. (see http://edtechpost.wikispaces.com/PLE +Diagrams#warlick). We encourage readers to visit the collection of diagrams to review a variety of PLE representations.

When a PLE has been conceived as a technical system or tool, it has often been described as a collection of several subsystems in the form of a desktop application or Web-based services (van Harmelen, 2008). Schaffert & Hilzensauer (2008) defined a PLE as a collection of social software applications the learner has collected that are useful for his or her own specific needs. Lubensky (2006) sees a PLE as a facility accessed by learners where content is organized and vetted for one's own learning needs. Downes (2006) is similar in his view that PLEs are Web 2.0 in their read-write ability but that they should probably be seen as a way for learners to access a large collection of applications and a network of peer learners. PLE pioneer Scott Wilson of CETIS defined the PLE as the collection of tools used in one's personal working and learning routine (Wilson, Liber, Johnson, Beauvoir, Sharples, & Milligan, 2006). The PLE involves using a combination of existing devices, applications, and services within what may be thought of as the practice of personal learning using technology.

PLE Examples

In a comprehensive Educause research bulletin on PLEs, Niall Sclater (2008) identified three perspectives on what PLEs should consist of and how they should function. The first perspective is that the PLE should be client software that mediates between the learner and whatever resources the learner wants or requires. The second perspective is that a Web-based portal can be an effective PLE without the need for client software. The third perspective is that PLEs are already here in the form of physical and electronic resources that learners can manipulate and customize to learn effectively (Sclater, 2008). Following is a brief summary of tools that, from the above three perspectives, can function as all or part of a PLE.

Client-based PLE tools

PLEX (http://www.reload.ac.uk/plex/) is an open source PLE prototype application developed at the University of Bolton. PLEX allows the user to seek out learning opportunities and manage them. PLEX supports standards such as RSS, Atom, and FOAF.

Colloquia (http://www.colloquia.net/) is a software application developed for group work. Once installed on each user's computer, Colloquia allows a user to create workgroups based on contexts or projects. These contexts allow for the sharing of resources, messaging, and project management. Colloquia was released as version 1.3 in September of 2001 and transitioned to open source in September of 2002. Colloquia is described as a conversation-based PLE (van Harmelen, 2006).

Web-based tools with PLE characteristics

Elgg (http://www.elgg.org/) is an open source social networking platform and e-portfolio tool. Elgg is server-based, meaning one can download, install, and host an instance of Elgg.

Chandler (http://chandlerproject.org/) is a server-based, open source personal organizer with calendaring and task management, and consists of a desktop application, Web application, and a free sharing and back-up service. Chandler was built for productivity as opposed to learning, but has some PLE characteristics.

EyeOS (http://www.eyeos.org) is an open source operating system that resides within one's web browser. So, one's files, applications, and settings are available at any networked computer.

Facebook (http://facebook.com) is a proprietary, Web-based, social networking platform, but has enough components and flexibility to be considered a form of PLE, even though it was not built primarily as a learning tool. Facebook includes a somewhat open API, extensibility, file sharing, forums, microblogging, instant messaging, and RSS feeds.

43 Things (http://www.43things.com) is a Web-based service where users post lists of resolutions or life goals they wish to accomplish. Users can find others with shared goals and form an ad hoc community for encouragement and accountability along the way. Many of the posted goals involve learning in some way.

Netvibes (http://www.netvibes.com) is a Web portal where users can personalize pages. Individuals can assemble favourite widgets, websites, blogs, e-mail accounts, social networks, search engines, instant messengers, photos, videos, podcasts, and more, all in one place. Netvibes is primarily an information gathering service, but one can see in this service the semblance of a PLE.

Two other examples described in reports include a model for an interactive logbook PLE (Chan, Corlett, Sharples, & Ting, 2005) and a "personal learning planner" (Havelock, Gibson, & Sherry, 2006). These are a few examples of tools that could be considered part of one's PLE — highlighted here to show possibilities or precursors of a construct being formed.

Why use a PLE?

We know that the majority of what a person learns will occur outside of formal instruction (Cross, 2007). A PLE can be seen as manifestation of a learner's informal learning processes via the Web. Learners have always depended on the support of their peers and peer networks to facilitate learning. In the physical world, these peer networks are experienced as lunchtime discussions, student organizations, communities of practice, brown-bag sessions, and study groups. What was lacking until recently was a way to effectively approximate these informal learning opportunities online. With recent developments in social networking, the Web is now a more people-oriented place rather than just an expansive information repository.

The PLE approach to online learning is buoyed by two factors. First, it mirrors what is happening in learners' "real lives" in terms of using myriad tools and processes for social networking and connectedness. Second, learners may have experienced limitations with what we call institutionally centred learning environments, embodied by learning management systems (LMSs).[1] While LMSs have served universities well in tracking students and orchestrating online courses ("learning management"), the learner is left with a less than optimal environment. It may not be in the learner's best interest to be "managed," but rather to be guided and encouraged. The comparison between a PLE and an LMS is presented in a later section of this chapter.

The central line of reasoning for the use of PLEs is the value of learner-centred instruction. One's stance on the importance of PLEs may rest on how one perceives informal learning and constructivist philosophy. Both informal learning and constructivism have the learner as the primary actor in knowledge building. The clearest argument for the PLE is that it allows the learners themselves to construct their own learning environments by forming communities, and creating, remixing, and sharing resources (Attwell, 2006).

Attwell cites the massive uptake of MySpace contrasted with the limited interactions via an institutionally controlled LMS as evidence that educational technologies have not kept pace with today's learners. Attwell posits that the predominant focus on "managing" via the institutional LMS has not resonated with modern learners, and that the educational system is in danger of being perceived as irrelevant or as an imposition (Attwell, 2006).

In an extensive report on PLEs, researchers with the Centre for Educational Technology and Interoperability Standards (CETIS) derived the following principles when examining current learning technologies (JISC-CETIS, 2007).

> Learning opportunities should be accessible to students, irrespective of the constraints of time and place.
> Learning opportunities should be available continually over the period of an individual's life.
> Effective teaching should have as its central concern the individual learning needs and capabilities of a student.
> The social component of learning should be prioritized through the provision of effective communication tools.
> Barriers to learning, whether they are institutional, technical, or pedagogical, should be removed.

In a similar report, Johnson et al. (2006) identified five major themes as a critique of current learning environments:

> desire for great personal ownership of technology;
> desire for more effective ways to manage technological services;
> desire for the integration of technological activity across all aspects of life;
> removal of barriers to the use of tools and services; and
> desire to facilitate peer-based working.

It is apparent from the conceptual definitions and the examples cited that the PLE is a response to the limitations of current learning environments as described in these reports. Following is a comparison of the LMS and the PLE.

PLE Compared with LMS

A LMS is a software application that has existed in some format since the 1990s in academia as well as in industry. Learning institutions as well as companies began to adopt the LMS in order to deliver instructional content and to control access to it. Corporations commonly use an LMS to track and report employee training completion and to deliver mandatory compliance training when necessary (Avgeriou, Papasalouros, & Retalis, 2003). Higher education has experienced a dramatic uptake in LMS use in recent years, and LMS use is now moving into secondary education (virtual high schools, etc.) as well. Following is a summary of LMS characteristics.

> LMSs concentrate on the course context.
> All resources are loaded and linked within the overall structure of a course.
> LMSs have an inherent asymmetric relationship between instructor and learner in terms of control of the learning experience.
> The learner's role is one of passive acceptance of content and the limited permissions set by the LMS.
> Every learner experiences content exactly the same way. Each learner interacts with content in an identical fashion.

Compliance with standards such as SCORM and IMS has caused LMS design to further solidify. LMSs are built on access control and rights (permissions) management, and only approved users can access the system. Finally, the scope of operation of the LMS is usually restricted to a single institution (Wilson et al., 2006).

There are certainly limitations to the current institutional approach to online learning (chapters 3, 6, 10). The LMS is not open to activities occurring outside its realm. The modern learner is steeped in an online environment of free-flowing content and interaction, is learning to navigate its complexity, and may view the institutional LMS as limited or inferior (Sclater, 2008). Researchers have identified from the literature these perceived failures of current online learning environments:

> Accessibility has only partially been achieved by moving the medium of dissemination onto the Web. However, barriers to accessibility remain, in the form of institutional procedures and usability.
> Institutionalization of learning technology creates an additional barrier through a milieu of interface constructs putting extraneous burdens on learners who must navigate between these systems.
> Current pedagogical practice is still teacher-centric. The promise of e-learning in enabling effective management of a diverse student population has only seldom been realized. At its worst, the VLE can be characterized as a giant photocopier.
> The process of education is primarily institution-centric, rather than learner-centric. (JISC-CETIS, 2007)

Scott Wilson et al. (2006) examined the design of LMSs and the alternative design presented by PLEs. The researchers compared LMSs to standards such as the VHS videotape and the QWERTY keyboard, and proposed that the LMS had become the de facto standard in online learning.

Unlike LMSs, PLEs attempt to manage the relationship between the learner and various Web-based services. PLEs do not attempt to integrate all the tools within one environment but rather to facilitate the sharing of content. The relationship of the learner with the PLE is a symmetric one in which the learner can produce and receive information within the same system. PLEs focus on facilitating connections using whatever standards are required. Finally, the scope of PLEs is global, in that there are no limitations to the PLE's reach via the Internet.

A PLE brings with it a host of changes for the learner, the institution, and the content. Anderson (2006) details several advantages of the PLE over the traditional LMS. With PLEs, the learner has a sense of self or identity beyond the classroom. As they direct their own learning, learners control the environment in which they work. The learner personally organizes the environment instead of operating within an environment that makes sense to the instructor or institution. The learner has responsibility for his or her own content. No longer a passive consumer, the learner is now in an ownership role. The learner's reach extends much farther than the traditional classroom and LMS. While taking part in various online communities of practice, the learner develops an online personality (Anderson, 2006).

Schaffert and Hilzensauer (2008) identified how facets of online learning differ in an LMS compared to a PLE, particularly in terms of the role of the learner, personalization, the social component, content ownership, organizational culture, and technical issues. Schaffert and Hilzensauer outlined clear challenges that learners will face when shifting to PLEs as a learning medium. Learners will be required to effectively select and review learning content independently; use several tools at once in a combination; understand the strengths of various Web 2.0 applications and services; have a better appreciation for intellectual property and ownership; and be internally motivated to learn.

Because technologies associated with the PLE are evolving, the PLE may become more advantageous over time. The accommodating nature of the PLE to new tools and services makes it difficult for LMS developers and vendors to keep pace. However, there are instances of current LMSs employing tools of the Web 2.0 evolution, such as chat, blogs, and wikis. The tension arises, however, in that these Web 2.0 tools are outward manifestations of an underlying ethos of social learning, communities of practice, and open resources (Downes, 2005). For example, some LMSs offer student blogs, but the blogs may not be accessible to readers outside the LMS. While an LMS can include Web 2.0 elements to its systems, it is rooted in the traditional instructor-centric model of instruction. Curricula are determined, courses are designed, networks extend only to the boundaries of the institution, and participation is limited to students paying tuition, and often only to the students in a particular course (see chapter 6).

The emergence of PLEs is less about establishing a new path in online learning than it is a response to the limitations of current online offerings. PLEs are not creating a market, but rather addressing an already apparent state of affairs. In a PLE, the learner is not restricted to only institutionally approved groups and resources. The PLE becomes the gateway to the Web where learners evaluate resources and make meaning of content. Learners are free to join any networks that make sense to them and offer value. This type of activity aligns with the concept of communities of practice (Wenger, 1998). We contend that communities of practice have more potential to be realized with the PLE than with the LMS. Wilson et al. (2006) summarize this effectively:

> The VLE is by no means dead, and those with investments in this technology will attempt to co-opt new developments into the design in order to prolong its usefulness. It is however the view of the author that the key distinctions between the VLE and the PLE are of a more conceptual nature than purely of features, and that ultimately alternatives such as the PLE model will develop in sophistication, making the VLE a less attractive option, particularly as we move into a world of lifelong, life-wide, informal and work-based learning.

Challenges to PLE Implementation

While the case for PLEs might be argued in educational technology circles, there are significant challenges to PLE success. These challenges are both technical and social. The majority of technical challenges involve how PLEs will integrate with institutional LMSs. We do think that institutional LMSs will exist long into the future. Therefore, the question is how will PLEs operate effectively within and outside the boundaries of institutional LMSs.

PLEs are challenged by the sheer scope of the online world. While LMSs provide demarcations between approved users and the outside, online communities can contain many thousands of participants and resources. Wilson et al. (2006) contend that emerging PLE technology might solve the issue of limitless resources by facilitating local filtering within a learner's PLE. In effect, trusted persons and processes become the "personal librarians" for the learner, mining through mountains of information and directing the learner to valuable resources (Martindale, 2007). We can see instances of this now with tools such as blogrolls and RSS readers. Users can construct and share lists of who they are reading (blogrolls) and what they are reading (RSS feeds). Microblogging tools such as Twitter (http://twitter.com) show whom a user is following and who is following the user.

The technical hurdles for PLEs can be considerable depending on one's definition of a PLE. A PLE as "a loosely joined combination of software applications aligned with a single learner to support specific needs" poses fewer technical challenges than does "a single application that can share data with all possible social software formats and e-learning applications." While a PLE can be a loose collection of social networking software, better utility and ease of use would come from tighter integration of these applications. Because PLEs are generally comprised of several social software applications, the skills necessary to manage all of these applications are considerable. The rate at which Web 2.0 applications arrive, expand, and sometimes disappear creates a challenge to learners looking for new components for their PLEs. Successful PLE learners must be able to navigate multiple interfaces, passwords, and content formats to benefit from the myriad offerings on the Web. Sclater (2008) describes the daunting task of simultaneously juggling multiple learning contexts and interfaces.

Any new system makes demands upon the user. Indeed, this user experience is common with any new tool or gadget. Each new tool represents what might be compared to a new grammatical rule to learn. Therefore, a multiplicity of tools represents an increase in complexity on the user. The user must manage this complexity, but the more tools a user has, the more difficult the management becomes. Not only must users learn new interfaces each time a new component is incorporated, but they must also learn how that new component interoperates with existing tools. PLE learners are required to spend higher proportions of their time learning and re-learning user interfaces of emerging Web 2.0 personal technologies (JISC-CETIS, 2007).

Johnson et al. (2006) write about the cognitive burden on the modern PLE learner faced with so many interfaces:

> An institution-controlled tool presents the user with a fixed interface of controls (instruments) that the user must learn to use effectively if they are to access the service provided. It is a feature of the current Web environment that the use of a large number of these interfaces creates an obstructive user experience, made worse by the lack of flexibility the user has for integrating the different services they access. To operate within this environment, the user must manage a number of different dispositions and skills required for different interfaces.

Moving away from the tightly controlled environment of an LMS with a clear delineation between expert and learner, the informal online learner is faced with the challenge of the constant evaluation of resources. Schaffert and Hilzensauer (2008) contend that there is a need for media-literate learners for the proper administration of these PLEs:

> [T]he change from content that was developed by expert and/or teachers towards possibilities and challenges to make use of the bazaar of learning opportunities and content leads to the necessity of advanced self-organising and searching in the Web — in other words: media competent learners.

Sclater raises a number of PLE implementation issues. System interoperability between LMSs and PLEs might be considered a utopian vision due to the business interests of LMS vendors. Why would a LMS vendor allow a PLE client to access the LMS functions without the user directly using the LMS? One must also examine the underlying assumption that learners are prepared to be responsible for managing their own learning environment and content. And there are questions about how the PLE reconciles with the traditional elements of formal education, such as syllabi, assignments, grades, and schedules. And the PLE "movement" at this point lacks a recognized charismatic leader or champion to push the development of PLE standards (while successful open source initiatives such as Apache and Linux did have recognized leaders [Sclater, 2008]).

Emerging technologies struggle to coexist alongside (and sometimes replace) current dominant technologies. There are three scenarios in which PLEs could coexist with LMSs. The first scenario would be the PLE existing in a "parallel life," dominating the informal learning space, while the LMS continues to dominate formal education. The second scenario would see LMSs gradually open their structures to include interoperability with PLEs. The third scenario would be the LMS attempting to co-opt elements of the PLE. This last scenario would likely reduce the transformative power of the PLE (Wilson et al., 2006).

Future Directions

Attwell (2006) writes that PLEs should operate online and offline, work on multiple devices, allow granular permissions control, support multiple learning contexts, be open to multiple sources, provide powerful searches, be easily updated, be easily installed and maintained, be extensible, provide multiple presentation options, have built-in interoperability, be based on standards, and help learners sequence their own content. With this as a checklist, clearly there is much work to be done for the PLE to be realized. Attwell (2006) and Sclater (2008) both comment on the relatively slow uptake of Web 2.0 technologies in formal education, which limits the trajectory of PLE growth.

For the PLE to gain ground in educational practice, instructor-centred instruction would have to become less dominant (Schaffert & Hilzensauer, 2008). While technology might enable better PLEs in the future, key development would be higher education institutions and corporate training departments fully embracing learner-centred learning. Attwell (2006) states that twenty-first–century industry will require employees to have ever-increasing technical competence to stay competitive. Modern workers will, by necessity, practise lifelong learning and take control of their learning processes. As learning becomes multi-episodic, the PLE will play a role in aiding modern learners.

There are a number of technologies and initiatives in development that could affect the PLE concept. For instance:

> The e-Framework for Education and Research (http://e-framework. org) is an attempt to create standards of interoperability for LMSs and related tools.
> Google's Open Social (http://code.google.com/apis/opensocial/) is a set of common APIs (application program interfaces) for building social applications across many websites.
> The Open ID project (http://openid.net/) is a shared identity project that allows Internet users to log on to many different websites using a single username and password (an identity).
> Moodle (http://moodle.org/) is a free, open source LMS that has the potential to be more learner-centred than the typical LMS.
> The Open Courseware Consortium (http://ocwconsortium.org/) is a collaboration of over 200 institutions that share open learning resources.
> The Mash-up Personal Learning Enrivonment, or MUPPLE (http://www.icamp.eu/watchwork/interoperability/mash-up-ples/) is a wide-ranging approach to PLEs focusing on the over-arching methods for creating an interoperable framework for different social networking applications and services.

Clearly there are issues facing the PLE, and a number of directions for future research and development. In terms of directions for research, we need a better understanding of how various social software applications are best used for learning; the implications of decentralized learning environments for institutions such as universities; the implications of learners being responsible for their own environments; how to maintain identity and manage privacy across multiple sites and services; and how PLEs can work alongside and integrate with institutional LMSs.

NOTE

1 Note that LMSs are often referred to as virtual learning environments (VLE), particularly outside of North America. In this chapter we use the terms LMS and VLE interchangeably.

REFERENCES

Allen, C. (2004, 14 October). Tracing the evolution of social software. Message posted to http://www.lifewithalacrity.com/2004/10/tracing_the_evo.html

Anderson, T. (2006, 8 June). PLEs versus LMS: are PLEs ready for prime time? Message posted to http://terrya.edublogs.org/2006/01/09/ples-versus-LMS-are-ples-ready-for-prime-time/

Attwell, G. (2006, 12 December). Personal learning environments. Message posted to http://www.knownet.com/writing/weblogs/Graham_Attwell/entries/6521819364

Avgeriou, P., Papasalouros, A., & Retalis, S. (2003). Towards a pattern language for learning management systems. *Educational Technology & Society, 6*(2), 11–24.

Chan, T., Corlett, D., Sharples, M., & Ting, J. (2005). Developing interactive logbook: a personal learning environment. *IEEE International Workshop on Wireless and Mobile Technologies in Education.* 73–75.

Cross, J. (2007). *Informal Learning: Rediscovering the Natural Pathways that Inspire Innovation and Performance.* San Francisco: Pfeiffer/Wiley.

Downes, S. (2005). E-learning 2.0. *eLearn Magazine.* Retrieved from http://www.elearnmag.org/subpage.cfm?section=articles&article=29-1

Downes, S. (2006). Learning networks and connective knowledge. *Instructional Technology Forum.* Retrieved from http://it.coe.uga.edu/itforum/paper92/paper92.html

Havelock, B., Gibson, D., & Sherry, L. (2006). The personal learning planner: Collaboration through online learning and publication. *Computers in the Schools, 23*(3/4), 55–70.

History of personal learning environments. (2008). *Wikipedia*. Retrieved 1 December 2008, from http://en.wikipedia.org/w/index.php?title=History_of_personal_learning_environments&oldid=251987270

JISC-CETIS. (2007). The personal learning environment: A report on the JISC CETIS PLE project. Retrieved 1 December 2008, from http://wiki.cetis.ac.uk/Ple/Report

Leslie, S. (2008, 4 June). A collection of PLE diagrams. Retrieved from http://edtechpost.wikispaces.com/PLE+Diagrams

Lubensky, R. (2006). The present and future of personal learning environments (PLE). Message posted to http://www.deliberations.com.au/2006/12/present-and-future-of-personal-learning.html

Martindale, T. (2007). Assembling your own personal learning environment. Presented for the Institute for Intelligent Systems Cognitive Science Seminar, University of Memphis, Memphis, TN.

O'Reilly, T. (2005). What Is Web 2.0?: Design Patterns and Business Models for the Next Generation of Software. Message posted to http://www.oreillynet.com/pub/a/oreilly/tim/news/2005/09/30/what-is-web-20.html

Schaffert, S., & Hilzensauer, W. (2008). On the way towards personal learning environments: Seven crucial aspects. *eLearning Papers*, (9).

Sclater, N. (2008). *Web 2.0, Personal Learning Environments and the Future of Learning Management Systems* (Research Bulletin, Issue 13). Boulder, CO: EDUCAUSE Center for Applied Research.

van Harmelen, M. (2006). *Personal learning environments*. Paper presented at the Sixth International Conference on Advanced Learning Technologies (ICALT '06), Kerkrade, The Netherlands.

van Harmelen, M. (2008). Personal learning environments. Retrieved 10 December 2008, from http://octette.cs.man.ac.uk/jitt/index.php/Personal_Learning_Environments

Wenger, E. (1998). *Communities of Practice: Learning, Meaning, and Identity*. Cambridge, UK; New York, NY: Cambridge University Press.

Wilson, B., & Lowry, M. (2000). Constructivist learning on the Web. *New Directions for Adult and Continuing Education, 88*.

Wilson, S. (2005, 4 October). Architecture of virtual spaces and the future of VLEs. Message posted to http://zope.cetis.ac.uk/members/scott/blogview?entry=20051004162747

Wilson, S., Liber, O., Johnson, M., Beauvoir, P., Sharples, P., & Milligan, C. (2006). Personal learning environments: Challenging the dominant design of educational systems. In E. Tomadaki & P. Scott (Eds.), *Innovative Approaches for Learning and Knowledge Sharing, EC-TEL 2006* (pp. 173–182).

LEARNING, DESIGN, AND EMERGENCE:

10

Two Case Studies of Moodle in Distance Education

> *Andrew Whitworth & Angela Benson*

Abstract

Course management systems (CMSs) display tendencies towards emergence, evolving through activity that takes place in many micro-level contexts. However, some systems are designed in more directive ways than others, and importantly, this is not just a factor of the type of CMS, but of the sociotechnical structures that exist around it. Directive systems increase the tendency that ways of working will be *reified* in the system, which then isolates it from organizational learning processes and blocks true emergence. On the other hand, responsive systems can act as a "boundary object" for multiple stakeholders, and can also broker the exchange of learning between activity systems in different universities. As an open source system, Moodle has the potential to be responsive, and we examine two case studies of its use in distance education. Our conclusion is that these program teams have succeeded in bringing their micro-level learning processes to bear on the central Moodle kernel, but not their host institutions.

Introduction

Between 2005 and 2007, our "Technology at the Planning Table" (TPT) project conducted eight qualitative case studies of distance learning programs across five universities in the UK and U.S. Our cases used a variety of course management systems (CMSs), including commercial, open access/open source, homegrown, and ad hoc (academic-created) systems. We concentrate here on lessons learned — by us as researchers

195

and by our subjects — regarding the use of open access/open source CMSs in constructing and delivering distance learning programs.

Although it is not only open source CMSs that can be emergent, we suggest that due to the way they are designed, the practices that contribute to their evolution are more likely to be inclusive and participatory. Operational proximity (Tagliaventi & Mattarelli, 2006) between technical support and teaching staff is easier to achieve with open source systems, and is of significant importance in improving the responsiveness of any CMS. We believe that these key concepts — participation, emergence, operational proximity, and responsiveness — increase the possibility that learning, from diverse professional and organizational perspectives, can actively contribute to the evolution of distance education teams and their CMSs. However, distance educators must consequently bear in mind that the CMS, and the organizational structures that surround it, will also be affected by the needs of other on-campus systems.

Emergence Through Activity

The TPT project uses the dynamic and holistic method of modelling activity developed by *activity theory* (Engeström, Miettinen, & Punamäki, 1999; cf. Bedny & Harris, 2005). Activity systems are comprised of relationships — and tensions — between many elements. Even systems built around the same basic technology, such as a CMS, will have diverse configurations of elements such as rules, divisions of labour, and external relationships (Benson, Lawler & Whitworth, 2008), which require that technologies, and organizations that create and use them, are both context-dependent and in a constant state of evolution.

De Wolf and Holvoet (2005) state:

> A system exhibits emergence when there are coherent emergents
> at the macro-level that dynamically arise from the interactions
> between the parts at the micro-level. (p. 3)

CMSs are emergent because the interactions that form them take place in many micro-level contexts. But the ways in which they *cohere* depend on the organizational structures with which they co-evolve (Andrews & Haythornthwaite, 2007). What micro-level contexts within a system — such as a higher education institution (HEI) — are permitted to influence macro-level outcomes? Organizational structures such as hierarchy and/or strict divisions of labour, will result in emergent systems that are less inclusive. These structures promote not participation in macro-level processes, but *direction* of some parts of the system by others.

In Benson and Whitworth (2007), we suggested that the alternative to a directive system was one that was *responsive*. We characterized both of the case studies that used commercial CMSs and one of the "homegrown" CMSs as directive; the rest, including the two cases with open source CMSs, we characterized as responsive. Responsive CMSs could respond to user needs innately, through flexible design, but also by being placed in a wider activity system that promoted active and direct negotiations between users and developers. On the other hand, directive CMSs could evolve in response to user needs indirectly, at best. They tended to represent higher-level control of the teaching and learning process: a system that had been largely shaped by decisions that were not inclusive, characterized also by a lack of "operational proximity" (Tagliaventi & Mattarelli, 2006) between technical support and teaching staff. The influence of operational proximity on communication within the system shows that responsiveness can be designed into, and emerge from, a sociotechnical system, and is not simply a characteristic of CMS type (commercial, open source, homegrown, or ad hoc) alone.

Participation and Reification

Emergent systems are highly complex and dynamic, throwing up organizational problems that are ill structured (Kitchener & King, 1990) and ambiguous (March & Olsen, 1979). In such circumstances, participation is an ongoing and social process of learning. Hence the development of *communities of practice* (Wenger, 1998), in which actors within a context develop their own understanding of it through the sharing of practices. This takes place largely informally, and is not directed by management decree or procedure. Indeed, it is often in opposition to them (see the example of the insurance claims processors in ibid. and in our examples below). Communities of practice can exist in any organization, but frequently develop "under the radar," adapting to, and possibly subverting or avoiding, policies mandated from above (ibid.). For instance, they may find "workarounds," or simply not implement decisions made by the centre.

Ambiguity is common in the education sector (March & Olsen, 1979). As a result, the professional practice of educators is not adequately promoted by centralized regimes of training, but is made more effective by participating in ongoing individual and group processes of self-reflection (Carr & Kemmis, 1986); of *learning how to learn* and becoming "reflective practitioners" (Schön, 1995). Such learning is facilitated by strong communities of practice (Friedman, 2001). Consequently, these communities are more significant and overt in educational organizations than in other sectors, except perhaps in other "professional" organizations (Mintzberg, 1989, p. 173–95). This helps explain the historic and decentralized structure of the typical Higher Education Institution, with strong community and network ties within disciplines, but only weak ties across them. What ties different communities of practice into a single HEI is *not* their core professional activities (teaching/learning, research, and professional development), but administration, and HEIs are "loosely coupled" (Weick, 1976) as a result. Within smaller educational communities, however, the level of participation and learning can, in principle, be high.

Wenger contrasts participation with *reification*:

> Where participation is about acting, interacting, and living in the world, reification is about the development (process and product) of artefacts and objects that embody aspects of the practice. Reification involves making aspects of the practice tangible, what Wenger calls giving "thingness" to the often implicit qualities of the practice. (Stuckey & Barab, 2007, p. 447)

Significantly, reification can happen at each end of the top–bottom organizational scale. Values, organizational goals, and procedures can be centralized, and thus reified, into *technostructures* (Mintzberg, 1989): parts of an organization mandated to design and control the work of others, such as business process analysts. Through reification, ways of thinking are "pushed" at members of the organization, locking activity in place around assumptions that become unquestioned and "natural" (Blaug, 2007; Whitworth, 2009, ch. 9). Though HEIs are historically decentralized, tendencies towards centralization and thus reification have increased following the widespread integration of ICT into many of their core activities (Robins & Webster, 2002), and the consequent strengthening of HEIs' technostructures. These are the "tangible" artefacts and objects that reify certain aspects of organizational practice.However, reification can also happen when communities of practice isolate themselves, drawing together around shared ties but excluding input from outside the community, turning "core competencies into core rigidities" (Brown & Duguid 1998, p. 97). Loose coupling between different parts of an organization makes innovations (the result of social learning processes) difficult to diffuse across community boundaries (chapter 11). This is one reason why Mavin and Cavaleri (2004) called academia "the last place to find organizational learning" (p. 287).

Both forms of reification — centralization and isolation — are often a response to the other. For example, where Bennett and Bennett (2003) say that "despite the increased pressure being placed on faculty to integrate technology in their courses, many are reluctant to do so" (p. 54), the *despite* might be better as *because of*. Communities may respond to increasing centralization with subversion or avoidance of new procedure. Their community-level solutions and workarounds become the object of their activity, rather than the CMS as a whole (Benson & Whitworth, 2007, p. 87–89). This is no more likely than centralization to lead to the questioning of basic assumptions held by communities of practice and thus locked into technological artefacts. Both organizational learning and professional development will likely suffer. A directive CMS can drive a "wedge" between communities of practice and the technostructure, encouraging each to reify its existing practice and thus retard critique, organizational learning, and the evolution of both the system and the practices it embodies.

On the other hand, a CMS that is negotiated between both the centre and the periphery can be an *architecture of participation* (see Garnett & Ecclesfield, 2008), promoting both professional practice and organizational learning. This would help the system to remain truly emergent: that is, emerging from the broadest range of micro-level contexts, rather than having its nature determined by only a limited subset of stakeholders. For this to happen, ongoing processes of *negotiation* (Cervero & Wilson, 1998) are required between various stakeholder communities, which challenge "the limits of each [stakeholder] community's beliefs" (Brown & Duguid, 1998, p. 98). Such negotiation is more likely to take place in informal work settings "on the ground" than in formalized meetings, and is facilitated by the existence of operational proximity between different stakeholder groups: that is, opportunities for them to *work together in a shared context* (Tagliaventi & Mattarelli, 2006).

How, then, can architectures of participation be facilitated in HEIs, in ways that work *with* both their loosely coupled structure and the new ICTs and external pressures; and that do not, as a result, encourage the reification and thus perpetuation of current practice, both at the centre and periphery?

Open Source CMSs as Boundary Objects

Embedding values into technology is how organizations learn: "through the storage of individual knowledge in organizational structure and routines" (Tagliaventi & Mattarelli, 2006, p. 293). But damaging reification occurs when different cognitive cultures (Whitworth, 2007) that could potentially contribute to a system design are no longer communicating across their boundaries. What becomes embedded will then be a singular perspective, that of an isolated community of practice (which might be core — the managers', for instance — or peripheral).

However, a truly negotiated CMS becomes a *boundary object*. Fischer and Ostwald (2005) say that boundary objects have meaning within the conceptual knowledge systems of at least two communities of practice. The meaning need not be the same — in fact, the differences in meaning are what lead to the creation of new knowledge. (p. 224)

They go on to say that

Boundaries are the locus of the production of new knowledge.… boundary objects should be conceptualized as evolving artifacts that become understandable and meaningful as they are used, discussed, and refined…

The interaction around a boundary object is what creates and communicates knowledge, not the object itself. Humans serving as knowledge brokers can play important roles to bridge boundaries that exist across or within communities. (Fischer & Ostwald, 2005, 224–5)

When multiple perspectives contribute to a boundary object, it becomes the locus of a *community of interest*. Fischer and Ostwald (2005, p. 213–4) suggest that these communities of interest address "the challenges of collaborative design involving stakeholders from different practices and backgrounds"; promote "constructive interactions among multiple knowledge systems"; and rely "on boundary objects to mediate knowledge communication." Crucial to this process is "the educational impact of participation itself" (Blaug, 2007, p. 41). A negotiated, participatory, and responsive CMS brings together the various cognitive cultures in an HEI (at both centre and periphery) within the boundary object that is the CMS.

As we have said (above, and in Benson and Whitworth, 2007), responsiveness in a CMS is not solely a property of open source technologies such as Moodle. In addition, it would be quite possible for a Moodle solution to be imposed from the centre and direct the behaviour of users, thus acting as a "wedge" between core and periphery. Nevertheless, the open source approach to technology development does provide certain channels for participation that other types of CMSs do not.

Many Moodles exist throughout the education sector. Moodle was specifically designed to be easy to adapt to different contexts (Dougiamas & Taylor, 2003), and it scales easily from single, one-off uses on a particular course to serving the needs of large universities. Also, in principle, *any* user can design a Moodle-based innovation that could be accepted into the central technological architecture, the Moodle kernel. Therefore, as well as being a boundary object at the organizational level, the Moodle.org community works at the meta-level to develop a shared understanding about the *architecture* on which local Moodles are based. This is, partly, a technical, programming task. But it is also a matter of developing shared understandings about the pedagogical (or other) principles that drive the technology. Moodle is based on social constructionist principles (Dougiamas & Taylor, 2003; Moodle.org, 2008), though, importantly, "Moodle doesn't FORCE this style of behaviour, but this is what the designers believe that it is best at supporting" (Moodle.org, 2008). In theory, through the "free market" principles of open source software, these principles are being constantly validated and dynamically updated by a global community of users. Although Moodle.org therefore exists to reify practices into the technological object that is the Moodle kernel, this reification is under constant review (chapter 1). In principle then, operational proximity is easier to design into, and be retained by, activity systems that use Moodle (or other open source CMSs) compared to other types.

In practice, however, Moodle is susceptible to distortions that affect any community that "focuses heavily on building a body of quality resources" (Stuckey & Barab, 2007, p. 446); "the 'grab and run' action of many new members becomes counter-productive to dialogue" (ibid.). Moodle could be passively consumed by users rather than being actively generated by them (see Luckin et al., 2010). This places the burden of development on only a small proportion of users. It is also a form of exclusion and isolation of practice. Also, work at the community of practice level will also be subject to distortions that originate outside the activity system, for example, pressures placed on course teams by technostructures and management at the institutional level.

A Tale of Two Moodle Sites

Our research included two program sites where Moodle was the CMS of choice. PAP ("Public Administration Programme") is a wholly on-line UK Masters program. It originated and was funded as part of the UKeU project and survived that institution's collapse (Conole, Carusi, & de Laat, n.d.). E-TECH is a wholly online US Masters program in education. The program originated with funding from the Sloan Foundation. The programs were very similar in organizational structure but very different in philosophies of online teaching and learning (see also Benson et al., 2008).

Program goals

Two primary goals drove the E-TECH program: 1) to provide a site for research into online learning tools, technologies, and strategies; and 2) to provide a stable and effective online E-TECH program. PAP's primary goal was to provide a stable and effective online program that was self-supporting.

Program and campus technology

E-TECH's selection of the open-source Moodle software as its course management system is reflective of the program's goal to be a research bed where instructor researchers could perform trials and demonstrate online technology tools and strategies. PAP's selection of Moodle was more practical. They had to quickly move from the vanishing UKeU platform, and Moodle was a reasonable alternative that was available on a local server.

E-TECH used Moodle and several other commercial and open source supporting technology tools in its courses, while PAP was a strict user of Moodle-only tools. Both the PAP and E-TECH campuses adopted Blackboard as the campus-wide commercial course management system. PAP's university did so despite PAP staff lobbying for Moodle. After this decision, the PAP program was directed to move PAP to Blackboard. PAP staff had to make a case for why they shouldn't move to the new system. The process was contentious, but PAP was allowed to continue its use of Moodle, though not indefinitely.

E-TECH staff have not been directed to move E-TECH to the campus system. In fact, the campus office that administers external programs like E-TECH provides E-TECH with technical support for Moodle and the other technology tools the program uses. The research objective of the program and the researcher roles that instructors play may keep E-TECH shielded from such influence in the future.

Program cultures

Because of the two-fold objective of the E-TECH staff, the E-TECH philosophy tends towards an open and non-standardized course design. Instructors are encouraged to experiment in their course design, which results in students having drastically different experiences in each course in the program. E-TECH operates its own budget, using funds generated by student enrolment and subsidized by the academic department in which it is housed. Finally, E-TECH staff fully support Moodle.org and participate frequently in its forums.

The PAP culture tends toward standardization of course design and tutor practice with the use of compliance documents, such as course development guides, tutor contracts, and student guides. PAP sponsors a yearly conference for tutors to further enhance the community aspect. PAP operates its own budget, using funds generated by student enrolment and subsidized by the academic department in which it is housed. PAP also fully supports Moodle.org and submits each new feature it develops to Moodle.org for inclusion in the base Moodle product. However, this is not quite as inclusive a process as it is with E-TECH, as the next section will show.

Program communities

Several stakeholder groups participate in the development and ongoing administration of both programs, but the divisions of labour differ between each system (here, see also Benson et al., 2008). For example, in E-TECH, instructors and developers work together to provide course content and activities. E-TECH staff (teaching and development assistants) build the courses, and instructors teach them. E-TECH staff and developers serve as the first line of technology and administrative support for instructors and students. E-TECH also benefits from a university-level academic support organization, which works with them to provide advanced software support, including fixes and new feature development.

Likewise, several stakeholder groups participate in the development and ongoing administration of PAP, but the relationships are different. While PAP staff remain the builders of courses, content and activities are provided by content experts, and then tutors, full-time and part-time, teach the courses. PAP staff are the first line of technology and administrative support for tutors and students, but advanced software support is less integral to PAP than it is in (and around) E-TECH. An external contractor provides advanced software support, including software fixes, new feature development, and Moodle.org liaison for submitting locally developed features. The university's technical support staff only support the university's standard virtual learning environment, Blackboard (eLearning), not Moodle.

Lessons Learned
Summary

While E-TECH and PAP have similar organizational structures, their reasons for choosing an open-source CMS such as Moodle and their philosophies of using it are very different. Within the program, PAP tutors are directed to use Moodle in certain ways, whereas E-TECH's researchers and instructors have more freedom to explore alternatives if they feel these would be more pedagogically effective on their course. However, PAP has moved over time to a less directive stance vis-à-vis its tutors.

These differences point to a key feature of open source systems: they can be standardized for users who want standardization and they can be individualized for users who prefer customization. This feature sets open source systems apart from commercial systems.

No cost vs. different costs

Often people think of the open source option for course management systems as a free or low-cost alternative to the major commercial systems. While it is true that the source code may be free or less expensive, there are hidden costs associated with the use of open source course management systems. The biggest of these costs is technology support and administration. E-TECH employed a Moodle programmer and technology support staff, while PAP purchased a Moodle programming and technology support contract from an external provider. In addition, these programs require pedagogical expertise in online course design and delivery. These skills are not necessarily found in Moodle programmers or technical support, so additional pedagogical support staff are also needed.

However, although the operational proximity between instructors, developers, and Moodle itself was slightly less in PAP than E-TECH, both teams were *active* users of Moodle, not just passive consumers of its benefits. In both cases, these teams did succeed in having the results of their reflective practice — their learning about the system-in-use — embedded not only into their local Moodle but also into the Moodle kernel. Particularly for PAP, in which members of the course team had less freedom and fewer resources with which to experiment and innovate with alternative technologies, this was a way of stabilizing the system-in-use, rendering the team as a whole less vulnerable to updates to the system coming in from outside, that is, being imposed on them as a result of changes to the Moodle kernel developed elsewhere. Their reflective practice, therefore, has increased the knowledge base of the team as a whole, and embedded that knowledge, at least partly, into the technological architecture. Active use of the CMS, therefore, leads to a more negotiation-based, participatory, and responsive system, as opposed to a directive one.

Centralization vs. localization

One observation that can be made from the PAP and E-TECH programs' use of Moodle is the tension that exists between campus-level administrators and systems and program-level administrators and systems. This tension exists because campus-level administrators and program-level administrators have different primary goals. In both E-TECH and PAP, campus-level administrators were concerned about security and the integration of course management systems with other campus systems for registration, security, and grading. These were not the primary goals of either of the programs.

The tensions suggest a question that campus administrators must address: what is gained from the centralization of course management systems and their support as opposed to what is gained from de-centralization? There are no easy answers. Benson and Whitworth (2007) determined that centralized systems tended to be less responsive to their users at the program level than de-centralized systems managed locally by the programs themselves. As a result, program-level administrators tended to use subversion tactics — employing workarounds to address system shortcomings instead of working with campus-level staff to address them — when required to use campus-level systems. Examples of subversive tactics include using the centralized CMS as a front-end to the program courses, but providing the actual content directly on the Web or with locally managed external applications. As we noted above, this is an example of the workarounds becoming the object of activity rather than the CMS, and the learning that these course teams engage in is consequently not feeding back into the system. In situations where this "subversion" happens — which included all three of the directive systems we researched (Benson & Whitworth, 2007) — the system cannot be said to be truly emergent.

This did not happen so obviously with either of our Moodle case studies. Both were self-contained in technological terms, and both expressed a commitment to a management style that they self-termed "laissez-faire" (E-TECH's course director) and "inclusive … enabling the people who work on the team to have as much responsibilty and as much ownership as possible for their work" (PAP's course director). E-TECH's director continued:

> You bring your best ideas in for your course, and we'll help you mix and match and merge that with the best ideas from technology, and we'll get the course up. And if you wanna ask some questions of us, we're there to help you. But we're not there to pass muster on your ideas, [your] pedagogical and course information ideas.

E-TECH's policy is facilitated by a research student who is also paid to act as the local Moodle developer, and as noted above, he has an active relationship with the kernel and Moodle.org. There is thus an ongoing process of negotiation occurring here, not only between members of the E-TECH team, but through this *brokerage* (see Fischer & Ostwald, 2005, p. 225, and above), E-TECH and other activity systems that share its technological architecture. For E-TECH, Moodle is a genuine boundary object working at both the micro-level and the wider macro-level structure. Though divisions of labour are stronger in PAP, this is at least in part explained by its courses being targeted at civil servants, rather than at educational technologists. Deliberate policy decisions were taken to standardize certain practices, as it was believed this would make the technology easier to use for its students. Teaching staff were also not expected to engage with CMS technology at the level of research and active use. Nevertheless, over time, a more participatory system is emerging at the micro scale, and Moodle has always been a boundary object between PAP and other systems.

Ideally, campus-level administrators must be sensitive to the different types of CMS users. Users who are delivering full programs online have different needs than users who are supplementing their traditional campus courses with online content, activities, and resources. The campus-level administrators on the E-TECH campus were sensitive to the needs of the program and supported the open source system. The campus-level administrators on the PAP campus were also sensitive to program needs, but they felt the campus security needs overrode those needs. As noted above, however, PAP has been able to defend itself from the top-down directives to change. Indeed, as a result of the case made by the staff, the campus-level e-learning administrator has requested certain changes be made to the Blackboard system before PAP's host institution fully adopts it. The investments made in learning about the technology have, in this case, been able to change practices in other parts of this loosely coupled HEI, albeit indirectly.

We suggest that one way campus-level administrators can address the centralization-decentralization question for fully online programs is to centralize the course management function but decentralize the technical support. By definition, open source systems can be responsive to user needs, but that responsiveness requires a strong set of technology skills and a high level of knowledge of the systems' features and processes. Unless this knowledge and skill sets are made available locally to the online program, the system will not be fully utilized by the program or made fully compatible with the program's needs. This corresponds to Tagliaventi and Mattarelli's (2006) suggestion that *operational proximity* is most helpful for facilitating the transfer of knowledge and innovation between different stakeholder groups.

Standardization vs. individualization

PAP and E-TECH adopted different philosophies for course design and delivery. The operating practices of the PAP staff yield a structured and controlled online course environment in which students face a consistent interface and operation in each module in the course. As noted above, since students are not technology experts and courses are not technology-related, this standardization is a positive characteristic of the program. There is, though, a downside to this standardization: it severely limits tutor decision-making when teaching a course. Thus, even though the PAP use of Moodle was responsive (Benson & Whitworth, 2007), standardization in course design limits that responsiveness at the tutor level. The PAP staff have recognized this unintended consequence and is working towards loosening some of the course standards.

E-TECH's course design philosophy, on the other hand, is that course design should reflect the interests and preferences of teaching faculty, yielding a set of courses with designs that vary by course and instructor. This philosophy is effective in E-TECH since the program's content is related to teaching with technology, so the students are enriched by the variety of course designs. The philosophy may not be appropriate, though, for programs where the content is not related to technology use. In those cases, the philosophy could become a hindrance to student learning.

Online program administrators would be better served by staking out a middle position along the standardization-individualization continuum, since neither PAP's extreme standardization nor E-TECH's extreme individualization is ideal (see chapter 3). A better solution would be one that balances the need for instructor flexibility in meeting course objectives with the student need for a non-intrusive use of technology. Once again, this is an example of how negotiation, participation, and responsiveness could be designed into an activity system.

Conclusion

Open source course management systems present the appearance of a low-cost, flexible solution to online course delivery, but that appearance is deceiving. The cost of the required programming and technical support must be added to the low cost of the source code. The inherent ability to customize an open source system for a particular use must be balanced with the need to provide students with an interface that does not detract from their learning. Finally, the ease of acquisition of open source systems by programs within institutions challenges the economies of scale that many institutions gain with centralized systems. Campus-level concerns can lead to distance educators being directed towards solutions that are less appropriate for their specific contexts.

In both our case studies, however, *learning* processes were taking place that were facilitated by the design of both the CMS itself and the sociotechnical activity system that surrounded the technology. Both case studies were differently configured, but both configurations were clearly the result of conscious design decisions made by program managers and (in E-TECH's case only) campus-level administrators. Operational proximity helped create "knowledge brokers," who were able to feed the reflective practices of course team members back into an emergent system. However, in each case, this was more apparent vis-à-vis Moodle itself than vis-à-vis each program's host institution. Though these examples show that loose coupling does not necessarily have to lead to "bottom-up" reification by isolationist communities of practice, they do suggest that it remains easier to develop communities of interest between *different* HEIs than within a single one.

Stuckey and Barab (2007) write that

> community design is never final: it requires a commitment to
> ongoing and sustained design, and management focus should be
> on community as a negotiation process. (p. 442)

Our research has led us to believe that to truly address the issue of organizational learning within HEIs, such a commitment is required *both* from management and the communities of practice, and is easier to sustain with a system that is responsive. Distance learning course teams should be aware that the responsiveness within their system is not, however, a given. It can be designed in, as a factor of management style, but it may also be challenged from without, or could decay, if not continuously refreshed by professional practice. The result may be a more directive system that ultimately could retard both the teams' and their host institutions' ability to learn about, and adapt to, the changes wrought by emergent technologies. Investing in operational proximity, which can both create knowledge brokers and boundary objects, and thus increase the knowledge base of the team as a whole, may be a significant investment for distance learning teams wishing to maintain their autonomy in the face of campus-level concerns.

REFERENCES

Andrews, R., & Haythornthwaite, C. (2007). Introduction to e-learning research. In R. Andrews & C. Haythornthwaite (Eds.), *The Sage Handbook of E-learning Research* (pp. 1–52). London: Sage.

Bedny, G., & Harris, S.R. (2005). The systemic-structural theory of activity: Applications to the study of human work. *Mind, Culture, and Activity, 12*(2), 128–147.

Bennett, J., & Bennett, L. (2003). A review of factors that influence the diffusion of innovation when structuring a faculty training program. *Internet and Higher Education, 6*(1), 53–63.

Benson, A.D., & Whitworth, A. (2007). Technology at the planning table: Activity theory, negotiation, and course management systems. *Journal of Organisational Transformation and Social Change, 4*(1), 65–82.

Benson, A., Lawler, C., & Whitworth, A. (2008). Rules, roles, and tools: Activity theory and the comparative study of e-learning. *British Journal of Educational Technology, 39*(3), 456–467.

Blaug, R. (2007). Cognition in a hierarchy. *Contemporary Political Theory, 6*(1), 24–44.

Brown, J.S., & Duguid, P. (1998). Organizing knowledge. *California Management Review, 40*(3), 90–111.

Carr, W., & Kemmis, S. (1986). *Becoming Critical: Knowing through Action Research.* Geelong, AUS: Deakin University Press.

Cervero, R., & Wilson, A. (1998). *Working the Planning Table: the Political Practice of Adult Education.* San Francisco, CA: Jossey-Bass.

Conole, G., Carusi, A., & de Laat, M. (n.d.). *Learning from the UKeU Experience* [e-Learning Research Centre working paper]. Retrieved 7 December 2008, from http://www.elrc.ac.uk/download/publications/ICEpaper.pdf

de Wolf, T., & Holvoet, T. (2005). Emergence vs. self-organisation: Different concepts but promising when combined. In Brückner, S.A. (Ed.), *Engineering Self-Organising Systems: Methodologies and Applications.* New York, NY: Springer.

Dougiamas, M., & Taylor, P. (2003). Moodle: Using learning communities to create an open source course management system. Paper presented at EDMEDIA 2003. Retrieved 28 November 2008, from http://dougiamas.com/writing/edmedia2003

Engeström, Y., Miettinen, R., & Punamäki, R.-L. (Eds). (1999). *Perspectives on Activity Theory.* Cambridge: Cambridge University Press.

Fischer, G., & Ostwald, J. (2005). Knowledge communication in design communities. In Bromme, R., Hesse, F., & Spada, H. (Eds.), *Barriers and Biases in Computer-Mediated Knowledge Communication.* New York, NY: Springer.

Friedman, V.J. (2001). Action science: Creating communities of inquiry in communities of practice. In Reason, P. & Bradbury, H. (Eds)., *Handbook of Action Research* (pp. 159–170). London: Sage.

Garnett, F., & Ecclesfield, N. (2008). Developing an organisational architecture of participation. *British Journal of Educational Technology, 39*(3), 468–474.

Kitchener, K., & King, P. (1990). The reflective judgment model: Transforming assumptions about knowing. In Mezirow, J. (Ed.), *Fostering Critical Reflection in Adulthood: A Guide to Transformative and Emancipatory Learning.* San Francisco, CA: Jossey-Bass.

Luckin, R., Clark, W., Garnett, F., Whitworth, A., Akass, J., Cook, K., Day, P., Ecclesfield, N., Hamilton, T., & Robertson, J. (2010). Learner generated contexts: A framework to support the effective use of technology to support learning. In Lee, M.J.W. & McLoughlin, C. (Eds.), *Web 2.0-Based E-Learning: Applying Social Informatics for Tertiary Teaching.* Hershey, PA: IGI Global.

March, J.G., & Olsen, J.P. (1979). *Ambiguity and Choice in Organizations.* Bergen, Norway: Universitetsforlaget.

Mavin, S., & Cavaleri, S. (2004). Viewing learning organizations through a social learning lens. *The Learning Organization, 11*(3), 285–289.

Mintzberg, H. (1989). *Mintzberg on Management: Inside our Strange World of Organizations*. London: Macmillan.

Ostwald, J. (1996). Knowledge Construction in Software Development: The Evolving Artifact Approach [Doctoral dissertation]. Boulder, CO: University of Colorado at Boulder.

Philosophy. (2008). *Moodle.org*. Retrieved 8 December 2008, from http://docs.moodle.org/en/Philosophy

Robins, K., & Webster, F. (Eds.). (2002). *The Virtual University? Knowledge, Markets, and Management*. Oxford: Oxford University Press.

Schön, D. (1995). *The Reflective Practitioner: How Professionals Think in Action*, Aldershot, UK: Ashgate.

Stuckey, B., & Barab, S. (2007). New conceptions for community design. In R. Andrews & C. Haythornthwaite (Eds.), *The Sage Handbook of E-learning Research* (pp. 439–465). London, Sage.

Tagliaventi, M., & Mattarelli, E. (2006). The role of networks of practice, value sharing, and operational proximity in knowledge flows between professional groups. *Human Relations, 59*(3), 291–319.

Weick, K.E. (1976). Educational organizations as loosely coupled systems. *Administrative Science Quarterly, 21*(1), 1–19.

Wenger, E. (1998). *Communities of Practice: Learning, Meaning, and Identity*. Cambridge: Cambridge University Press.

Whitworth, A. (2007). Researching the cognitive cultures of e-learning. In R. Andrews and C. Haythornthwaite (Eds.), *The Sage Handbook of E-learning Research* (pp. 202–220). London: Sage.

Whitworth, A. (2009). *Information Obesity*. Oxford: Chandos.

11 INSTITUTIONAL IMPLEMENTATION OF WIKIS IN HIGHER EDUCATION:

The Case of the Open University of Israel (OUI)

> *Hagit Meishar-Tal, Yoav Yair, & Edna Tal-Elhasid*

Abstract

This chapter reviews the experience gained at the Open University of Israel (OUI) in implementing wikis in its academic courses. The first part discusses the strategy that has been employed to support the implementation of wikis in learning and teaching, concentrating on three perspectives: the technological, the pedagogical, and the administrative . The second part assesses the implementation process in terms of sustainability and diffusion. The experience gained at the OUI and the model of implementation developed by its leading team could serve as a model for the implementation of additional new and emerging learning technologies in distance learning institutions.

Introduction

Implementing a new e-learning technology in higher education institutions is a complicated process. Most of the literature discusses the implementation of various e-learning tools, often collectively known as Learning Management Systems (LMSs) or Virtual Learning Environments (VLEs). These environments usually contain course materials and forums for asynchronous online discussions. The focus of previous studies was on the transformation that was required from the university in moving from traditional face-to-face teaching to online teaching (Garrison & Anderson, 2003; Goodyear, Salmon, Spector, Steeples, & Tickner, 2001; Hegarty, Penman, Nichols, Brown, Hayden, Gower, Kelly, & Moore, 2005; Nichols & Anderson, 2005). The focus of the present work is slightly different: the change we discuss here is the transformation required with the adoption of wikis into distance teaching and learning.

Wikis are collaborative writing tools (Augar, Raitman, & Zhou 2004; Bruns & Humphreys, 2005; Lamb, 2004). As such, they are not suitable for use in all courses, but only in courses in which collaborative learning may be effective. Nevertheless, they possess a revolutionary potential in terms of pedagogy. The change required of the university in this respect is quite dramatic, moving from individual distance Web-assisted self-learning to online collaborative group learning.

The OUI Background

The Open University of Israel (OUI) is a distance learning university, established in 1974, which presently enrolls more than 40,000 students. In Israel, it is considered the pioneer and leader in implementing innovative e-learning technologies to support and enhance learning. The university was the first in Israel (1996) to develop an online learning platform (named OPUS) of its own. Each course in the OUI has a website to which the course staff can upload learning materials such as briefs, presentations, and enrichment material. Each course has its own message boards and discussion groups, to which messages are relayed by the course team, and where the students can post questions and queries, which the course staff or their peers can answer and discuss.

The adoption of e-learning technologies is dictated by the pedagogical goals of the university and the students' needs. As a distance learning university, OUI students can benefit from using an online learning environment, which allows them better contact with the academic staff, better access to learning materials, and an opportunity to collaborate from a distance (Harasim, 2000; Hiltz, 1990).

Before the Internet era, the OUI model of teaching was based exclusively on the self-study model. Students received a course kit containing printed and audio-visual material, which was sent to their homes by (regular) mail. They had to submit some written assignments to their tutors during the semester and pass the final exam at the end of the course. Some students found it very difficult to adjust to this model of self-study. They felt lonely and isolated, and often dropped out during the first semester.

OPUS enables the emergence of innovative teaching methods relying on a more social and constructive approach to distance learning. By using the Internet, students can communicate and collaborate with each other and with their tutors from a distance through conferencing tools and discussion groups (Garrison & Anderson, 2003; Salmon, 2005). They can discuss issues online and learn through debate, and they can expose their work to peers and get their advice and feedback. They can assess other students' work and comment on it. They can learn collaboratively in groups although they cannot meet face to face, and they can be part of a learning community (Hiltz, 1998). For this mode of collaboration, we regularly employed Internet tools that were not originally designed for this type of activity. Thus, the advent of wiki technology offered us an opportunity to use its unique aspects in a distance learning setting.

The implementation of wikis at the OUI started in September 2005. MediaWiki software was adapted to the university's LMS. During the first stage, a pilot group was set up, in which six courses were given the opportunity of using the platform and developing course assignments for their students. This pilot study was essentially a feasibility study, which proved to be very successful in terms of student and staff satisfaction (Tal & Tal, 2006). During 2007 and early 2008, a set of actions was taken to fully implement and integrate wikis as one of the mainstream learning tools used in the university.

The Multi-Dimensional Implementation Strategy Framework

The "sustainable embedding" (Sharpe, Benefield, & Francis, 2006) of a new learning technology in the institution demands a complex set of changes and transformations. Its success depends on the willingness and capability of the academic staff to embrace the new technology, and on the ability of the institution to manage and coordinate the process of implementation using a holistic approach (Nichols, 2007).

It demands the development of a detailed, multi-dimensional institutional strategy to cover all aspects of implementation: technological, pedagogical, and organizational (Koper, 2004).

The technological aspect of implementing a new learning technology in the organization involves a whole range of technical issues: choosing the right software and hardware to meet the needs of the institution, the students and the academic staff; maintaining the technology; and supplying the end users with the proper support (Schonwald, 2003).

From a pedagogical perspective, the adoption of e-learning requires changes in teaching approaches. Teaching face-to-face or the traditional mode of distance teaching (correspondence courses) is different from teaching online. Special skills are required from the instructors to carry out online teaching (Goodyear et al., 2001). This is also true in the transformation of distance teaching from the self-learning model to collaborative learning, as in the case of the adoption of wikis into teaching and learning. Staff development is therefore a crucial component in the implementation process, as stated by Hegarty et al. (2006, p.1):

> Capability in e-learning was wider than just the acquisition of technical skills, and required staff development activities that would help staff overcome fear and anxiety, motivate them to become involved in new technologies for teaching, and develop a clear appreciation of pedagogy related to e-learning.

The third aspect to be considered in the implementation process is organizational. Nichols and Anderson (2005) claim that the strategic challenge to the institution is to "efficiently coordinate e-learning development without stifling innovation." Implementing learning technologies successfully depends on a set of institutional moves and conditions. From the administrative perspective, moving towards online learning requires full coordination between managers, administrators, and faculty (Koper, 2004; Nichols, 2007; Salmon, 2005).

From an economic perspective, it is a matter of massive investments (Salmon, 2005). New technologies should be evaluated by the institution and proved to be not only effective but cost-effective as well (ibid.).

Table 11.1 sums up the overall parameters involved in the implementation process:

Table 11.1 The multi-dimensional framework of the implementation of wikis

Technical aspects	Pedagogical aspects	Organizational aspects
Software adjustments	Dissemination initiatives	Rules and regulations
Hardware adjustments	Staff development	Reporting procedures
Technical support services	Pedagogical support	Financial aspects
	Assessment and evaluation	

The following paragraphs review in detail the actions taken by the OUI Center for Technology in Distance Education (SHOHAM) in implementing wikis.

Technical Aspects

Software adjustments. MediaWiki, which was chosen to be the platform for the wiki activities, is open source software that can be downloaded from the Internet and installed on the university's servers. The generic version of the software cannot meet the university's special requirements. Some adjustments had to be made before starting to use the software, such as defining a bureaucrat and sysops (managers), and designing the edit toolbar to make it more user-friendly and function-rich. Also, language and design issues had to be solved, since Hebrew is written from right to left.

MediaWiki is external software and as such, is not an integral part of the OUI's VLE. It was crucial to integrate it into the VLE by connecting it to the identification and authorization system that contains the database of the users and their passwords, which are needed to control access the wikis.

A PHP programmer was tasked with the management and maintenance of the wiki server and software. The job description also required improving the wiki and responding to new requirements. For example, MediaWiki did not have enough statistical reports, an essential tool for the course coordinators (academic staff) to control and assess the assignments and students' performance. Based on the requirements, an in-house statistical tool was developed that gave a real-time report on the overall assignment and its various components.

Hardware adjustments. In the early stages of the project, the Media-Wiki software was installed on a test server, which was unsuitable for use as a "Production Server." In the transformation from a pilot to an integrated tool in the university's arsenal, the system administrator had to provide a suitable server and backup system and move the wiki to it.

Technical support services. The OUI personnel of the Technical Support Center for students and staff were trained to be able to assist students by telephone in case of any difficulties in operating the wikis.

Pedagogical Aspects

Implementing wikis at the OUI was not only a technological endeavour; it was mainly a pedagogical revolution. The OUI existing model of teaching and learning depends mostly on self-study methods. We had to make some major changes and adjustments at the pedagogical level in order to generate the desired change.

Dissemination initiatives. The first challenge of the project managers was to identify the potential courses for carrying out wiki activities. At the end of the first pilot, a special seminar was held for a wide academic audience, presenting the pedagogical potential of wikis for different models of assignments. The successful activities from the pilot stage were discussed and ideas for continuation were presented. This seminar was planned and successfully served as a recruiting tool, and consequently additional course coordinators expressed their wish to join the project in the following semester. The wiki project was also reported in internal university publications and in the Israeli media (Haner, 2006; Shalev, 2006).

Staff development. Although MediaWiki is user-friendly software, training is required at both the technical and the pedagogical levels. A five-hour workshop was given to new course coordinators (CCs) who joined the project. During the technical part of the workshop, the CCs learned to operate a wiki both as a user and as the manager of the environment (sysop). They were exposed to the special tags of MediaWiki, they learned how to compare versions, and they learned how to look at a user contributions report. In the pedagogical part of the workshop, they were shown how to design a successful wiki assignment and how to run it during the semester. They received advice on how to encourage students to participate and to collaborate, how to assess these assignments, and how and what level of involvement they should invest in the assignment while it is being developed by the students.

The workshop was compulsory for CCs who wished to join the project, and at the end of the workshop the participating CCs were required to prepare outlines for their future wiki assignment.

Ongoing pedagogical support. SHOHAM staff continued to support the CCs during their first semester and onwards, assisting with technical and pedagogical issues on demand.

A special Web portal for the wiki project was published, containing important information on the educational wikis, such as links to published academic papers, printed tutorials on the usage of wikis, a set of audio-visual short clips that explain how to use wikis, the seminar recording, and the list of courses and CCs in the project, with links to their wikis. The portal also contains a training zone in which course coordinators who are interested in trying or practicing how to use wikis are able to do so (http://wiki-openu.openu.ac.il/courses/wikiop).

Organizational Aspects

Rules and regulations. Within the traditional model of teaching of the Open University, students are not allowed to submit assignment in pairs or in groups, only individually. If two assignments are identified as containing a high level of resemblance, they are not accepted and the students may be subjected to disciplinary measures.

Wiki assignments are, by definition, collaborative assignments. In order to move to a stage in which wiki assignments are not considered "experiments" but are part of the pedagogical options that can be utilized by the CCs, some modifications to the regular rules of the university had to be made. The university's academic committee had to decide whether a collaborative wiki assignment is acceptable and under which conditions.

The instructional designers and pedagogical experts in SHOHAM prepared a White Paper containing all the regulations that had to be implemented when working with wikis. This paper covered issues such as the maximum percentage of the final grade that could be given to the wiki assignment, the assignments' assessment procedures, the training of CCs, and the question of how to use wikis — as compulsory or as optional assignments. The OUI academic committee accepted most of the recommendations in July 2007, emphasizing that a wiki assignment should replace an existing "regular" one, and not be added to the total number of assignments in the course. Since the fall semester of 2008, the wiki collaborative assignments have been considered a legitimate option and are an established teaching tool. This development demonstrates the possibility of emerging technologies impacting, changing, and moulding the organization in which they are being deployed (chapter 1).

Financial aspects. Designing, implementing, and assessing a wiki assignment is a time-consuming activity. To achieve CCs' participation in and persistence with wikis, there must be a fair fee related to the work. The shift from a pilot project to a mainstream teaching tool required establishing criteria for payments to the CCs that are not wholly different from the reward system for normal assignments. Otherwise, large-scale usage of wikis with a higher level of payment would create a huge burden on the university's budget. Thus, two levels of wiki duration and accompanying fees were established: one for a short-term assignment (up to two weeks) and the other for a long-term assignment (more than two weeks and up to two months).

Furthermore, since access to computers and the Internet is still not considered compulsory for studying at the OUI, students who state that they are unable to take part in a wiki assignment should be offered an alternative task, equal in content and pedagogical value.

Reporting procedures. The assignment reporting procedure in the university is rigorous, both in its traditional format (assignments sent by mail) and in its online format (assignments submitted online). The tutors receive the assignment (hard copy or electronic), assess it, complete the feedback and grade, and send it back to the Center of Learning Achievements. In the case of a wiki assignment, this procedure is inadequate, since the wiki assignment is not submitted to the assessor, but takes place within the wiki website. The procedures and the online assignment reporting system had to be changed in order to enable reporting grades for wiki assignments.

Assessment and Evaluation

The wiki project was assessed and evaluated from the initial pilot stage in various ways:

(1) Each semester, a survey was distributed to the students asking about their satisfaction with learning on the wiki platform. The findings are used to improve the design and management of collaborative assignments.

One of the outcomes of the students' survey was the insight that students want more feedback and tutor involvement during the time allocated for completing the assignment.

(2) Each semester, the CCs had to submit a report that described their wiki assignments: the level of commitment (compulsory/optional), the type of activity (glossary, web query, etc.), and the level of

collaboration. The forms are gathered into a database and saved for further research.

(3) The log files on the wiki server are accessible to the academic staff, who are able to conduct their own research on wikis. Students in the MA program in educational technologies are also encouraged to conduct research on wikis using qualitative and quantitative methods.

(4) Every year a report is written and submitted to the university chancellor describing advancements in the wiki project.

Diffusion and Sustainability of Wikis in the OUI

The basic measurement of the successful implementation of e-learning technologies in institutions is adoption (Hegarty et al., 2006; Nichols & Anderson, 2005). Adoption can be measured by two parameters: diffusion and sustainability.

Diffusion is measured in terms of scope: how many users, courses, or faculties have adopted the new tool (Nichols, 2007; Rogers, 2003).

Sustainability is measured in terms of time and continuous use. The sustainable embedding of e-learning is indicated by the number of courses that make use of the technology an integral part of their course's learning environment for a long period of time (Sharpe et al., 2006).

While measuring the diffusion and sustainability of wikis in academic courses, one must bear in mind that wikis are not "core technologies," like VLE platforms, but only a "peripheral technology" (Salmon, 2005). They are not suitable for use in all courses, but only in courses in which collaborative learning may be effective. Therefore, the rate of implementation should not be measured against the total number of courses the university has to offer, but with reference to the courses that joined the project.

The figures below present the current state of wikis in the OUI in terms of diffusion and sustainability.

Figure 11.1 Number of wikis per semester

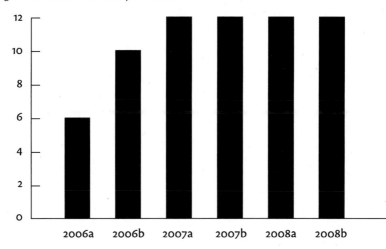

Figure 11.1 presents the growth of the usage of wikis in the OUI for the period 2006–2008 (every course could open only one wiki per semester). The overall picture shows a growth trend in 2006 and at the beginning of 2007, with stabilization during 2007–2008. This level of usage should be maintained over time, in order to achieve sustainability of the project.

Figure 11.2 Number of wikis and the number of courses per faculty (2006–2007)

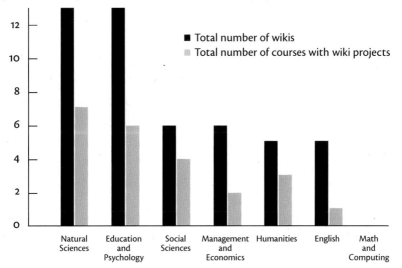

Figure 11.2 shows the distribution of wikis in the different academic departments. The black bars represent the number of wikis opened for the faculty, and the grey bars represent the number of courses that carried out a wiki assignment. The differences between the columns reflect the fact that some courses carried out a wiki activity more than once. This is a positive indication for the sustainability of the wiki project (see also Table 11.2).

The wiki was used mostly in the natural sciences and in the education and psychology departments, and was not used at all by the department of computer sciences and mathematics. The reason for this may stem from the lack of adequate mathematical capabilities within the wiki editing tools (equations cannot be easily written). The wiki editing tools should be improved before other potential users in additional disciplines are able to use it.

Table 11.2 Number of semesters of wiki usage per course

Number of semesters	Number of courses
1	9
2	10
3	2
4	3
5	2
6	1

Table 11.2 shows the number of semesters in which the courses operated a wiki. The longest term possible is six semesters. The shortest is only one semester. Not all the courses started in the same semester and could operate a wiki for all the six semesters. Out of nine courses that operated a wiki only once, six were taken for the first time during the 2008 spring semester (2008b), the last one in the present research. This means that only three courses (11 percent) dropped out of the project after one semester; the other courses that took part in the project found it useful and wished to continue using it for at least another semester. Only one course operated a wiki in all six semesters.

The wiki project is still in its early stages, and summaries and conclusions are naturally limited. However, the research on the diffusion and sustainability of the wikis in the OUI should continue as the project develops.

Problems and Obstacles

The wiki project in the OUI is successful in terms of diffusion and sustainability. Nevertheless, there are still some problems that must be considered.

(1) The wiki is not yet fully integrated into the VLE system. It is connected to the main VLE, yet works on another platform and with a different programming language. Therefore, maintaining the system is not easy and requires special effort from the system administrator.

(2) MediaWiki is not equipped with a graphic user interface, meaning the students have to be engaged in writing tags as well as content while working on the wiki. Some CCs find it not friendly enough and therefore refuse to take part in the project (see chapter 10).

(3) Collaborative writing is new to the OUI students, since they are used to studying alone. Although most of the students who participated in wikis did report a high level of satisfaction, they had numerous objections and fears at the beginning of the assignment. Some students (a minority) even refused to participate in the collaborative assignment, although it was compulsory.

Summary and Conclusions

The OUI's wiki project is a good example of the successful implementation of a new technology and innovative pedagogies in a higher education institution. It was flexible, quick, and required little prior set-up, often encountered during the development of new software. The project began with a small group of "wiki pioneers" (the typical innovators and early adopters, according to Rogers, 2003), and then extended into a large-scale project that eventually became an integrated part of the OUI arsenal of learning technologies. This process can serve as a model for the implementation of emerging technologies in other educational/

learning institutes. The model is based on six components: promotion, technical and pedagogical training, technical adjustments, institutional adjustment, assessment, and administrative arrangements.

A successful implementation must act in all the dimensions simultaneously in order to achieve good results, a quick and sustainable diffusion, and high levels of student achievement and satisfaction.

REFERENCES

Augar, N., Raitman R., & Zhou W. (2004). Teaching and learning online with wikis, *Proceedings of the 21st ASCILITE Conference, 2004.* Retrieved from http://www.ascilite.org.au/conferences/pertho4/procs/augar.html

Bruns, A., & Humphreys, S. (2005). Wikis in teaching and assessment: The M/Cyclopedia project. Retrieved from http://snurb.info/files/Wikis in Teaching And Assessment.pdf

Garrison, D.R., & Anderson, T. (2003). *E-Learning in the 21st Century: A Framework for Research and Practice.* London: RoutledgeFalmer.

Goodyear, P., Salmon, G., Spector, J.M., Steeples, C., & Tickner, S. (2001). Competences for online teaching: A special report. *Educational Technology Research and Development, 49*(1), 65–72.

Haner, L. (2006). The OUI started using wikis in 15 courses. *The Marker Online,* 23.8.06 (Hebrew). Retrieved 20 September 2007, from http://www.themarker.com/tmc/archive/arcSimplePrint.jhtml?ElementId=lh20060823_7741

Harasim, L. (2000). Shift happens online education as a new paradigm in learning. *The Internet and Higher Education, 3*(1–2), 41–61.

Hegarty B., Penman M., Coburn D., Kelly O., Brown C., Gower B., Sherson G., Moore M., & Suddaby G. (2006). eLearning adoption: Staff development and self efficacy. *ascilite 2006* Conference Proceeding, 3–6 December, Sydney, Australia. Retrieved from http://www.ascilite.org.au/conferences/sydney06/proceeding/onlineIndex.html

Hegarty, B., Penman, M., Nichols, M., Brown, C., Hayden, J., Gower, B., Kelly, O., & Moore, M. (2005). Approaches and implications of eLearning adoption on academic staff efficacy and working practice: an annotated bibliography. *New Zealand Ministry of Education.* Retrieved from http://cms.steo.govt.nz/NR/rdonlyres/89765CF4-A2ED-4088-9AE5-0097F7E7324C/0/ALETliteraturereview.pdf

Hiltz, S. (1990). Evaluating the virtual classroom. In Harasim, L. (Ed.), *Online Education.* New York: Praeger, 134–184.

Hiltz, S.R. (1998). Collaborative learning in asynchronous learning environments: Building learning communities. Paper presented at the WebNet 98 World

Conference of the WWW, Internet and Intranet Proceedings, Orlando, Florida.

Koper, R., (2004). Learning technologies in e-learning: An integrated domain model. In Jochems, W., Van Merrienboer, J., & Koper, R. (Eds.), *Integrated e-Learning: Implications for Pedagogy, Technology and Organization* (pp. 64–79). London: Routledge.

Lamb, B. (2004). Wide open spaces: Wikis, ready or not. *Educause Review*, 39(5), 36–48.

Nichols, M. (2007). Institutional perspectives: The challenges of e-learning diffusion. *British Journal of Educational Technology*, 39(4), 598–609.

Nichols, M., & Anderson, B. (2005). Strategic e-learning implementation, *Educational Technology & Society*, 8(4), 1–8.

Rogers, E.M. (2003). *Diffusion of Innovations*. New York, NY: Free Press.

Salmon, G. (2005). Strategic framework for e-learning and pedagogical innovation. *The Association for Learning Technology Journal*, 13(3), 201–217.

Schonwald, I. (2003). Sustainable implementation of e-learning as a change process at universities. *Paper presented at Online Educa 2003*, Berlin, Germany.

Shalev, E. (2006). WikiOP: The wiki portal of the OUI has been lunched. *Reload Magazine*, 29.8.06. (Hebrew). Retrieved 20 September 2007, from http://www.reload.co.il/article.asp?iobty=2846

Sharpe, R., Benefield, G., & Francis, R. (2006). Implementing a university e-learning strategy: Levers for change within academic schools. *The Association for Learning Technology Journal*, 14(2), 135–152.

Tal, H., & Tal, E. (2006). "Collaborative assignment in WIKI environment." Paper presented at Chais Conference for Research in Technology Learning, The Open University of Israel, Ra'anana, Israel (in Hebrew).

12 THE USE OF WEB ANALYTICS IN THE DESIGN AND EVALUATION OF DISTANCE EDUCATION

> *P. Clint Rogers, Mary R. McEwen, & SaraJoy Pond*

Abstract

One main challenge that has faced distance education since its inception has been a relative lack of knowledge concerning how students actually interact with the materials. This has made it difficult to decide if changes in content and/or methods make a positive or negative impact on learner behaviour and overall outcomes. Simply because students are at a distance, educators do not get the same kinds of immediate explicit and implicit feedback that comes when face to face. Web analytics provide an incredible opportunity for educators to receive helpful information regarding their students' usage and behaviour patterns — on a scale that has the potential to transform the entire industry. Utilized to date primarily in business to track the online behaviour of consumer groups and to test related marketing efforts, web analytics can also be used in distance education to improve the tracking of learner behaviour and to test the impact made by changes in content or presentation. In this chapter, we introduce web analytics, discuss its impact through a case study, and offer a vision of what impact data-driven decision making through the use of web analytics can make on distance education now and in the future.

Introduction

As distance education changes and grows, so too should the tools and techniques used to design and evaluate it. One of the major challenges facing those who are designing and evaluating distance education is to better understand if and how people actually utilize the educational experiences they create (Cadez, Heckerman, Meek, Smyth, & White, 2003). When do users access their educational materials? What type of browser, operating system, screen size, and connection speed are they using? How do they navigate through what is presented to them? How long do they take to complete certain activities? How does their behaviour relate to their results? Interestingly, as distance education becomes more Internet-based, data can be collected that helps answer these types of questions (Rogers, Flores, & Matthews, 2007). This chapter describes the potential for using web analytics in the design and evaluation of distance education. We describe how current web analytics technology can be used to track and understand the behaviour of students in order to improve their overall experience through a case study outlining the implementation and initial utilization of web analytics with one of the largest distance education providers in the U.S. We then suggest ways in which emerging technologies might enhance this understanding, and how distance educators can take an active role in shaping the future of web analytics.

According to the Web Analytics Association (2005), web analytics is defined as "the measurement, collection, analysis, and reporting of Internet data for the purpose of understanding and optimizing Web usage." Typically, web analytics have been used in marketing and business settings to monitor and test the best ways to drive traffic to a website, as well as to track user behaviour while on the website, in order to maximize the conversion of visitors into customers. Baker (2007) insightfully observes that "the rapidly growing volumes of computerized data has keyed the need for development of more automated ways of extracting actionable knowledge." Web analytics is one powerful way of extracting actionable knowledge in both business and in the field of distance education. We will begin by outlining the key terms and general process of utilizing web analytics. Next, we will use a case study to explain the practical implications in applying web analytics in distance education. We conclude with our vision of what the future may hold and with some questions for future research.

Web Analytics Terms and Processes

When adhering to strict ethical standards, many websites can currently use Internet-based tools to track the behaviour of users on their own site in order to improve the site and the experience for their users. Standard ethical practice includes not tracking a user when he or she is not on your site, as well as keeping all individual user data anonymous. Data of individual users is not considered nearly as valuable as the conglomeration of data from hundreds, thousands, and even millions of users. As data is collected and analyzed, trends are noted, hypotheses are formed, and alterations to the website based on those hypotheses can be implemented and tested.

For web analytics to be useful, organizations must first determine what types of outcomes they desire from users. There are so many metrics that *could* be tracked that it is absolutely essential for stakeholders to identify the metrics most meaningful to them — the ones they want tracked and monitored on a regular basis. These are called the Key Performance Indicators (KPIs). KPIs are determined from the Key Business Requirements (KBRs) or, in the case of distance education, the Key Educational Requirements (KERs). What are the main objectives that you want your website to accomplish? Which metrics will provide you the most meaningful information about how well you are accomplishing them? Take the dashboard of your car as an analogy. Many metrics could be tracked and reported about the current state of your car: combustion chamber temperature, fuel/oxygen mixture, fan belt RPM, and many more. But only a few are displayed on your dashboard, and most of us actively *use* only a selection of those. So, why are fuel level, speed, and turn signal functionality displayed on your dashboard while coolant level and spark plug efficiency are not? First, because these former metrics have direct, observable consequences to you, the operator (we all know how costly speeding tickets can be). Second, because you, the operator, can do something about them (the pressure of your foot on the accelerator directly affects your speed). Similarly, certain web analytics metrics become KPIs because of their impact on the ultimate outcome, as well as your ability to make actionable decisions based on them.

Depending on which vendor you decide to use (which will be discussed later), data is collected and reported in slightly different ways. Almost all web analytics providers, however, report the following types of data (although they may define them slightly differently):

> unique visitors (either from a unique IP address or an instance of a cookie),
> visits (valuable because each visit represents an opportunity for meeting the relevant KPI),
> page duration (as well as visit duration and bounce rates),
> pathing (including entry and exit pages), and
> visitor demographics (geographic location, time of day/week/month, technology used, etc.).

Data like this makes it fairly easy to discover which pages or sections of your site are the most popular and effective in helping users accomplish any objectives that can be tracked and measured. Data segmentation afforded by more sophisticated tools allows for more nuanced insights to arise regarding particular user groups. Additionally, options are even available for combining data from non-Internet sources with web analytics in order to infer relationships between web usage patterns and information from other user contact points (e.g., in a brick-and-mortar store or classroom).

With all the powerful metrics at your fingertips from even basic web analytics services, it is easy to feel overwhelmed by the amount of data. New users have often described their initial experience as "drowning in data" (Snibbe, 2006). It is also easy to lose track of what you are actually trying to discover, and what you would do with that information if you knew it. Here, it may be helpful to point out the critical yet often subtle distinction between *outputs* and *outcomes*: many analytics metrics simply (and valuably) track the *outputs* of our distance education efforts — how many people visited the site, where they came from, how long they stayed on a particular page, which of our expensive simulations they actually interacted with, their scores on our mini-assessments, and so on.

Interesting as they may be, many of these metrics become meaningful and actionable (and could therefore be termed KPIs) only when they are tied into the overall picture of *outcomes,* which often goes far beyond clicks and even conversions. What kind of penetration are we getting in our target demographic? What are learners actually taking away from the course, and how has that changed their behaviour?

In helping you understand some of the basics of using web analytics, we now will describe the initial implementation and analysis process of one distance education provider.

Case Study (BYU Independent Study)

Brigham Young University Independent Study (BYU IS) is based at a large, private, western university, and services approximately a hundred thousand distance education students in all fifty states and numerous other countries. Students were enrolled in more than five hundred high school or university courses. In 2008, several members of the BYU IS marketing team attended a presentation given by the primary author of this chapter (who was then faculty in the business school and teaching courses on web analytics). They soon brought him in as a consultant in order to figure out how best to utilize web analytics in their marketing efforts. This naturally led to utilizing web analytics in evaluating user behaviour in their online courses, as well. Some of the major decisions and outcomes of this process to date are described below.

Hiring or training expertise in Web analytics

Using web analytics can be a powerful way to make data-driven decisions, but data can also be overwhelming, confusing, and misleading at times. Having simply a surface-level understanding regarding web analytics is probably not sufficient in many implementation situations or when using the data to make significant, costly decisions. Common mistakes many beginners make can be learned from through painful experience, or avoided through hiring someone with previous experience.

BYU IS chose to hire a consultant who would provide expertise in the key initiation decisions, as well as provide training for selected staff (until most of the analysis activities could be performed internally). They also offered certain graduate students, who were being trained by the consultant, access to the data for research purposes in exchange for their analysis and resulting recommendations.

Determining KERs and KPIs

"Not everything that can be measured is important, and not everything that is important can be measured."
— Albert Einstein

The mission statement of BYU Independent Study is: "To make quality educational experiences available to all who can benefit from individualized learning." This statement was broken down into the following two KERs: (1) quality educational experiences, specifically those intended for (2) individualized learning. As a starting place, each of these KERs was assigned related KPIs. In a later section, we will describe those KPIs in relation to the data analysis of a single course.

Obviously not everything that is important can be tracked or measured using data collected from web analytics, so discussion also occurred regarding how to use other research methods to triangulate and enhance the data retrieved from web analytics. In many cases, one research source will provide valuable insights into data collected with a different method. The main goal in defining KERs and determining KPIs, however, is to clarify what indicators *can* be measured and then select those indicators that are most important to your objectives.

Choosing a vendor

As mentioned previously, there are a variety of vendors offering web analytics tools and services (for example, AWStats, Analog, Google Analytics/Urchin, Yahoo/IndexTools, Woopra, Microsoft/Gatineau Project, and Omniture). There are free software and services, and premium software and services. The premium options offer significantly more support and more options for customization, and allow you to find answers to more detailed questions, but the costs could be prohibitive for many distance education providers. Each vendor will offer you information regarding its service and you can decide which to choose in light of your particular needs and budget.

After some discussion, the initial decision of BYU IS was to use Google Analytics. This was based on ease of implementation and cost (free at the time of use). As BYU IS' questions and needs become more sophisticated, it is likely that they will upgrade to a premium service (Omniture, for example).

Implementation

In the case of Google Analytics, the basic code is fairly simple to insert in the relevant pages. Additional effort was needed, however, to implement code in a way that data could be collected and reported from groupings of courses into the same report suite (in addition to report suites for individual courses). For example, data can be viewed for all English vs. math and science courses, all university vs. high school courses, and so on. This enables analysts to see patterns more generally, and for decision makers to compare and contrast data from the relevant sections of courses. More detailed implementation would be needed to engage in A/B or multivariate testing (processes that allow users to compare the conversion results of two or more designs in a simultaneous randomized test).

Analysis

BYU Independent Study offers courses in over fifty subject areas, including Anthropology, Organizational Behaviour, Exercise Science, Slavic Languages, and Ancient Scripture. To begin exploring the potential of web analytics in distance education, one of the largest courses (a college algebra course) was chosen for initial analysis. The analysis in this section mainly applies to this course.

The graduate students who performed the analysis were concurrently enrolled in an introductory course on web analytics in online instructional design and evaluation. As part of a class assignment, they attended a presentation regarding the aims of BYU IS and were given access to the Google Analytics data. They were limited in some respects, as they did not have ready access to other data sources, and the data in Google Analytics had only been collected for the three or four months previous to their analysis. Table 12.1 captures the KERs, KPIs, and initial observations from the analysis.

Table 12.1 Summary analysis of college algebra course

KER	KPI	Initial Observations
Quality educational experiences	Material is engaging	> Average page views per visit are approximately equal to average pages per lesson. > As lesson number increases, average page views decrease.
	Educates students	> Page views and time on page metrics for the online evaluation are surprisingly low.
Individualized learning	Material is clear and easy to navigate	> A noticeable page-viewing trend is consistent throughout the course and across visitor segments. > Certain pages are viewed exclusively by a specified visitor segment.
	Anytime, anywhere	> Predictable viewing trends are observed by day of week and time of day. > The course has minimal international access, and majority of visits are from two states.

The first KPI that BYU IS chose to explore in connection with their KER of "quality educational experiences" was "student engagement with the course material." The debate about what constitutes "student engagement" and how it might be reliably measured extends to web analytics. Page-view metrics seem one of several reasonable factors to consider when attempting to measure this elusive construct. The analysts noticed a simple yet interesting correlation between page views per visit and pages per lesson. The averages of each of these metrics were quite close: average page views per visit = 6.5, and average pages per lesson = 5.09, suggesting that on average, a student's plan is to complete one lesson in a sitting. This is something to consider when determining the amount of content one lesson should cover. Armed with this data, designers and analysts could conduct systematic investigations into the effects of lesson length on page views, and even tie that data to quiz performance and other evaluative feedback to further optimize their content structure.

Further analysis of page-view metrics revealed several other interesting trends. First and most obvious was that with each subsequent lesson, the average unique pages viewed decreased, falling quite consistently for the initial eight lessons preceding the first midterm. The average unique page views then generally levelled off with only a very minimal decline afterwards (see Figure 12.1).

Figure 12.1 Average unique page views per lesson

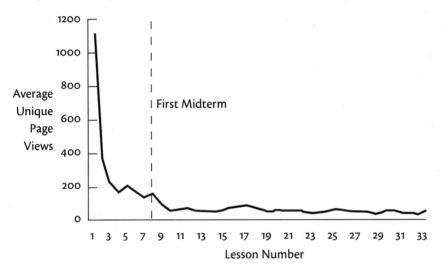

The data analyzed covered a four-month period from the first of July to the first of November. The observed decrease could have been due to attrition, or to students spending more time on their face-to-face classes and less on independent study courses, or to some other factor altogether. The analysts concluded that it would be easier to justify a particular hypothesis about the decline if there were a year of data to examine, which is the amount of time allotted to finish an independent study course without applying for an extension. Time-based trends like this one can be especially insightful when dealing with user experiences and conversion goals, like many in distance education, that extend over more than a single visit.

Determining the extent and value of the education students receive from a course is unilaterally difficult. This case study proved exceptionally so, as the analysts did not have access to any data about completion or scoring of assignments or midterms. Determining that the data from online course evaluations could prove an appropriate substitute, they analyzed page visits and time-on-page statistics for the course evaluation page. The results were quite surprising: there were only 26 visits from the 1,460 total visitors, and of those few visits, only 8 students spent more than one minute on the evaluation (which contained at least a couple dozen multiple-choice questions and some free-text questions). As it turns out, there was a good reason for this: a written evaluation is handed out with each proctored final. The inclusion of an online evaluation seemed to be a vestigial part of the course buried at the bottom of the menu, and any resulting data simply ignored. This analysis experience illustrates well the importance of critical questioning and communication in dealing with web analytics data. Had the team taken the data from the evaluation page at face value, without any further investigation as to possible causes, their analysis, recommendations, and any resulting adaptations to the site would have been irrelevant (if not an all-out waste).

An additional trend was found when unique page views were analyzed by lesson. The first page of each lesson had many more unique page views than any other page in the lesson. The unique page views fell sharply even between the first and second page of each lesson. Then they remained fairly constant, until the last page of the lesson where there was a very noticeable uptick in unique page views (see Figure 12.2 and Figure 12.3).

One probable explanation for the disproportionate number of first-page views is the navigational structure. Students first navigate by lesson number. Once the lesson is chosen, students are automatically taken to the first page of the lesson. It is only after lesson selection that the next navigational level (pages in the lesson) is displayed for selection. At that point a "course-wise" student who understands that these pages are not necessary for lesson completion may readily skip to the last page, where they are directed to a (third-party) off-site link to complete

Figure 12.2 UPV by lesson page for "All Students" segment

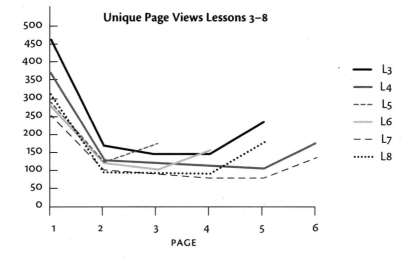

Figure 12.3 UPV by lesson page for "Engaged Students" segment

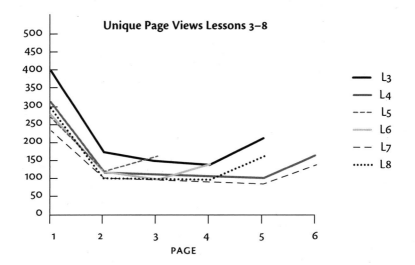

the lesson and assignment. This also could be a plausible explanation for the uptick in views for the last page of each lesson. Here, the team suggested a change that might improve ease of navigation as well as enabling better tracking of student intent: changing navigation of lessons and pages to naming according to content, rather than numbering alone. Often, substantive changes that improve site usability and support student learning can be double-purposed to provide richer analytics data that will help perpetuate the cycle.

A week or so before the presentation, Google rolled out a new analytics suite, including custom segmentation and custom reports. The analytics students took advantage of this, and after a little online research, decided to create two segments: an "all students" segment and an "engaged students" segment. To determine the criteria for inclusion in the "engaged" segment, they simply required the average page views and the average time-on-page to be greater than the average of those metrics for the "all students" segment. The most surprising results came in discovering that though the *numbers* changed for the two segments, the trends remained the same. These trends become visually apparent when the data is exported to Excel and displayed as line graphs, as seen in Figure 12.2 and Figure 12.3.

So if the trends remained the same, what differed between the two visitor segments? Data sorting and analysis revealed fifteen lesson pages that were viewed exclusively by the "engaged" segment. These pages were all middle pages (neither first nor last pages) of lessons towards the end of the thirty-three-lesson course (lessons 24, 28, 30, and 33). They were also all pages that were fairly complex in the worked examples and/or formulas covered. Is inclusion in the "engaged" segment a predictor of course completion, or is it an indication that the students are actually having a harder time digesting the material (thus spending more time on each page and viewing more pages)? To answer that question, more information and more analysis is needed.

If "anytime" and "anywhere" are important KPIs, then there should be some supportive evidence of that in the analytics data. The client was quite interested in the "where" and "when" data summary. Site access was reported by time of day and by day of week. Not surprisingly, the biggest day for site access was Monday, with Tuesday through Thursday access being fairly equal, and then falling off on Friday, with Saturday and Sunday being the days with the least traffic. As far as time of day: timing was split fairly evenly between 10 AM and 10 PM, with the hours of noon to 4 PM slightly heavier. It is important to note that often the metrics of initial interest to a client are not necessarily the ones that can yield the most relevant insights or have the most impact on the outcomes.

The geographic distribution of site visitors was also reported. By far the biggest percentage of all site users were from California and Utah. The Independent Study management confirmed that this aligns with registration data. It is interesting to note that although analytics for other college-level distance education courses showed a significant number of international visits, the international visits for the College Algebra course were minimal. It is counterintuitive to think that this could be solely a language translation issue, as mathematics could be considered a symbolic language in and of itself with few cultural or international barriers. Could a lack of international students (from Canada, Japan, China, Germany, and the UK) enrolled in the College Algebra course possibly indicate less of a need because post high-school students in those countries are more proficient in basic algebra? Though there is too little data to draw such a sweeping conclusion, this type of question points out the potential for web analytics to inform us about educational issues and trends that go well beyond the design and evaluation of a particular course.

Additional questions for future analysis include: What else can be determined given more in-depth access to course content and metrics, including third-party content, and test and completion data? Do these other sources of data confirm or refute web analytics data? Were navigation recommendations implemented? Where are the best sections to set up A/B testing to confirm/refute improvement?

Discussion

Effective use of web analytics data in distance education, and most other applications, comprises four basic objectives (Hendricks, Plantz, & Pritchard, 2008), perhaps more aptly termed opportunities:

> First, we must **define** the goals/objectives of the interaction. One of the unique challenges of using web analytics in distance education, as opposed to marketing or business applications, is determining how to translate Key Educational Requirements into Key Performance Indicators — essentially tracking learning through clicks. After all, learning the chemical processes of photosynthesis through an on-line simulation may involve the same physical behaviours (clicking, reading, scrolling, etc.) as purchasing a duvet cover, yet certainly the mental behaviours of these two interactions would be significantly different. The challenge for distance educators using web analytics is to discover how the learning and the cognitive processes and behaviours of their users are manifested in their online processes and behaviours.

> Secondly, we must **measure** both the outputs and outcomes of the interaction. Here's where web analytics services truly shine. Once the KPIs, conversion goals, and funnels are defined, the computer and the user do the rest. Still, no matter how specific we make our KPIs, no matter how precisely we define the funnels that lead to them, in the end we are still trying to get a picture of human learning through online behaviour, and we will always have cause to wonder at times whether we're really measuring what we think we're measuring, whether what we think we're measuring is even what we *should* be measuring. A little puzzling over those questions can be quite healthy.

> Thirdly (perhaps most the crucial objective), we must **use** the resulting data to make improvements in the interaction. A question to be asked is this: "What would I do with this information if I had it?" If you do not know how you are going to use the data and what changes you will make as a result of different possible outcomes, then you should consider exploring other metrics where there is some action-able outcome. Especially in web analytics, where so much data can be gathered with so little effort, distance education professionals must pay particular attention to utilization.

> Finally, analytics data thus interpreted and utilized may be **shared** to the benefit of users, other practitioners, and the distance education community as a whole. Web analytics in general, and especially its application to the field of technology-enhanced and distance education, is an emerging discipline. Valuable lessons that could help everyone include: how to incorporate analytics dashboards into the design workflow of distance education resources; how to translate learning objectives into effective KPIs; and how to create innovative metric mashups that combine metrics to illuminate deeper outcomes, deeper characteristics of the user, and their interactions. Insights like these could prove invaluable in bringing the power and agility of web analytics to education, and the depth and subtlety of education to web analytics.

These challenges are by no means reasons to shy away from the use of web analytics in the design and evaluation of distance education efforts. In fact, integrated web analytics data can help meld evaluation and design processes — and researchers and practitioners in distance education may in fact be uniquely positioned to take the use of analytics data in design process and strategic decision-making to a new level.

Conclusions

We conclude with our vision of what the future may hold, and some questions for future research.

Future vision and research questions

It is possible that the future of distance education could be dramatically influenced by information acquired through web analytics. The potential to collect and analyze real-time data from vast numbers of students could teach us a lot about how people interact with and learn in online learning environments. In reaching this potential, certain questions deserve more exploration:

> How do different segments of students (geography, age, gender, education level, major, etc.) interact with online resources?

> What are the common KPIs applicable to industry?
> What are the most effective ways to implement multivariate testing?
> How can web analytics be used to give individual students information regarding the relation of their own use patterns and results of others with similar patterns?
> What are the best ways to automate some decisions based upon data indices?
> How can the data gained through web analytics be combined with other evaluation methods (e.g., qualitative methods) in order to give a more complete picture of learner intent?
> How will data-driven decision-making through the use of web analytics change the processes by which distance education is designed and evaluated?

This chapter provided insight into the way in which web analytics might be used in the design and evaluation of distance education. Cadez, Heckerman, Meek, Smyth, and White (2003) propose that "arguably one of the great challenges ... in the coming century will be the understanding of human behaviour in the context of 'digital environments' such as the web" (p. 399). While acknowledging that using web analytics in distance education is an emerging endeavour, several strengths and opportunities are apparent. Behaviour and results in online environments can be monitored and analyzed with more ease and agility than ever before. While maintaining an emphasis on high-quality ethical standards, web analytics provides a clear opportunity for monumental contributions in making data-driven decisions in the design and evaluation of distance education. The capability to continually track and monitor learner behaviour on such a large scale, and the insights gained from doing so, could transform the way we think about distance education.

Note: Interested readers can find a list of additional readings and resources about web analytics in education at http://tinyurl.com/wa-resources.

REFERENCES

Baker, B.M. (2007). *A Conceptual Framework for Making Knowledge Actionable through Capital Formation* [Doctoral dissertation]. University of Maryland, College Park, MD: University College Press.

Cadez, I., Heckerman, D., Meek, C., Smyth, P., & White, S. (2003). Model-based clustering and visualization of navigation patterns on a website. *Data Mining and Knowledge Discovery, 7,* 399–424.

Hendricks, M., Plantz, M.C., & Pritchard, K.J. (2008). Measuring outcomes of United Way-funded programs: Expectations and reality. In J.G. Carman & K.A. Fredricks (Eds.), *Nonprofits and evaluation. New Directions for Evaluation, 119,* 13–35.

Rogers, P.C., Flores, D., Matthews, K. (2007, September). *The Learner/Teacher, Online Platforms, and Web Analytics in Learning Design.* Presented at the Open Education 2007: Localizing and Learning, Utah State, Logan, Utah.

Snibbe, A.C. (2006). Drowning in data. *Stanford Social Innovation Review,* Fall 2006, 39–45.

Web Analytics Association. (2010). *The Official WAA Definition of Web Analytics.* Retrieved 23 March 2010 from http://www.webanalyticsassociation.org/?page=aboutus

13 NEW COMMUNICATIONS OPTIONS:
A Renaissance in Videoconference Use

> *Richard Caladine, Trish Andrews, Belinda Tynan, Robyn Smyth, & Deborah Vale*

Abstract

Distance education has changed over the decades from a largely isolated, paper-based learning experience to one where rich visual and aural interactions with peers and teachers are now possible. Mitigating the effects of distance has been at the forefront of many who manage, design, teach, or learn with distance education. Video communications can in many ways address these effects. Internet Protocol (IP) video communications have become more relevant than ever as students and their teachers seek to interact with one another as they go about their learning. Renewed interest in communication tools has predominantly arisen due to increased access to the Internet, and on one level represents a renaissance. Further, there are environmental, technological, and economic drivers that will increase the use of internal-based video communications. However, many who teach or manage distance education do not have access to the knowledge and skills that make for effective and efficient use of video communications. A starting point is the discussion of issues and factors key to the scalability, sustainability, and pedagogical considerations of video communication in distance education contexts.

Introduction

It has been argued that the history of distance learning in higher education can be described as a series of generations, and that each new generation is defined by changes in practice and/or changes in technologies (Caladine, 2008a; Nipper, 1989; Taylor, 2001). In the first generation, known as the correspondence model, students received largely text-based materials via mail and had little, if any, interaction with one another or the teacher. In some cases, interactions with their peers and teachers were provided through blocks of intensive residential experiences, but these were not available for all courses and could not always be attended by all students. Perhaps some of the blame for the high attrition rates that characterized correspondence courses could be ascribed to this lack of opportunity to interact. Later generations of distance education were characterized by technologies that provided opportunities for interaction with content, in the form of audiotapes and videotapes, and later, learning management systems. Subsequently, possibilities for person-to-person interaction increased. Initially these technologies were audio technologies, such as two-way radios, audio-conferences, and audiographics. It is generally known that early experiences with videoconference were often characterized by users' dislike of being "seen" by a camera, and the experience of e-mail and other computer-mediated communications has been that being able to see participants is not a necessary prerequisite for effective and efficient communications. However, being able to see the reactions of others is an important aspect of interaction, and over the past twenty years or so, videoconferences have become increasingly popular for mediating distance. As the technology improves, there is a renaissance or renewed interest in using IP for videoconferences. Higher education has thus far been the biggest user of video communications worldwide, and this is predicted to continue and spread across the increasing variety of software solutions (Greenberg, 2008).

Drivers for the Uptake of Video Communications

Synchronous communications technologies, whether for personal use or for use in organizations or education, are at a watershed. While audio has been the default for many years, the use of video for two-way communications is increasing for several disparate, coincident, and substantive reasons.

Costs

A common form of visual communications, videoconference, became popular in distance education in the late 1980s even though the high costs of ISDN[1] often kept the bandwidth low, resulting in poor picture quality. However, many factors have caused a continued increase in videoconference use (Greenberg, 2008). In particular, these include the savings due to the change from ISDN to the Internet. Around the turn of the millennium, connections to the Internet became fast enough to support high-quality videoconferences, and the last few years have seen higher speeds that afford further increases in picture and sound quality. Today, high-definition cameras and screens are becoming the standard in videoconference technology, primarily due to the small price difference between standard and high-definition videoconference appliances. The change from ISDN to the Internet for video communications changed the cost structure with particular implications for education providers. ISDN lines were, in the main, owned by telecommunications companies that typically charged on the basis of cost per line and cost per kilometre or mile. Thus, the cost of ISDN videoconferences increased with the distance between the connected locations as well as the number of lines used to improve quality, often making it prohibitive for everyday teaching and learning activities. The cost structure of the Internet is independent of the distance between locations. Thus, for the same quality, a conference between adjacent buildings will typically incur the same network costs as one that connects countries on opposite sides of the world. Testing[2] in 2002 showed that a videoconference from Australia to the UK cost in the order of $1,200 AUD per hour using three ISDN lines and about $1.50 AUD per hour using IP. Further cost advantages are achieved by multi-campus institutions that own their networks. In these cases, once the infrastructure is in place, the network costs are those of maintenance.

Climate and Economic Change

Issues around global warming and climate change have encouraged many institutional leaders in education and commerce to rethink the necessity of business travel. Videoconference and its higher-quality counterpart, telepresence, provide more environmentally friendly alternatives that are also time-saving. As access to these technologies for students as well as academics and researchers is becoming easier, their uptake is set to rise. At almost the same time, the world has entered a period in which the economies of many countries are slowing or are in recession. It is reasonable to assume that resulting hard times will make many business personnel reconsider the costs of travelling to attend face-to-face meetings. Anecdotally, and at least in the authors' institutions, the trend is to cease or limit travel and opt for videoconference or telepresence.

Video is becoming the standard in interpersonal communications

For many years, telephony was the standard for synchronous communications between distant parties. However, the technology of traditional telephony is not as entrenched as once thought. Many organizations have made the change to Voice over Internet Protocol (VoIP), in which the Internet is used for telephonic communications. This has two distinct advantages. First, the cost structure of the Internet (as mentioned above) can deliver cheaper long-distance calls. Second, organizations can reduce their infrastructure costs by having only one network to install and maintain. Following immediately after Voice over IP is Video over IP, and the advertising material indicates that many VoIP providers are including video communications applications with VoIP solutions.

In addition to the use of Voice and Video over IP by organizations, the past five years have seen a marked rise in the use of VoIP and Video over IP for personal communications. A recent survey of university students (Caladine, 2008b) showed the use of video communications applications gaining on audio. Popular communications tools such as Windows Messenger and Skype, as well as video communications embedded in social software, were becoming popular. For example, Tokbox and Friendvox are video communications applications that can be embedded in Facebook. Further, the survey indicated that nearly 80 percent of the surveyed students used the applications for an average of 8.6 hours per week.

The year 2008 marked the fifth anniversary of Skype, one of the world's most popular IP communications technologies. When Skype was first launched, it was audio only. In the ensuing years, video communications became an optional extra. The latest release of Skype reveals a change in emphasis to a video communications tool that has the option be used as an audio communications tool.

Other recent changes that will create upward pressure on the usage levels of video communications are the proliferation of 3G mobile telephones and the transition of communications applications from audio to video. The recent release of the Apple iPhone in Australia was met with extremely high levels of uptake (as in other countries), and as 3G mobile phones are also Web browsers, they can be used as Video over IP devices. In Australia the number of mobile phones (or cell phones) has exceeded the total of the population, and 3G mobile phones outnumber older generations of this technology. Combinations of tools such as Personal Digital Assistants (PDAs) with 3G mobile capacity and installed applications are becoming attractive to university departments (e.g., medicine, education, engineering) where students are regularly on remote practicums and need to engage with fellow students and staff to complete required learning activities. The convenience of these tools will create additional pressure for high-speed wireless/mobile networking across Australia and elsewhere in the next five to ten years.

Challenges for Distance Education

These three categories of drivers indicate that video communications for organizational and personal use will increase, and is predicted to increase in higher education by 24 percent over the next few years to 2013 (Greenberg, 2008, p. 90). While this is good news for the suppliers of the technology and for the environment, it poses challenges for those engaged in distance education. These challenges will be economical, technical, and pedagogical. Economical and technical challenges will relate to the sustainability, scalability, and interoperability of applications and appliances. Pedagogical challenges will arise from the need to develop appropriate teaching and learning practices. A repeating dilemma will arise with each new wave of technology: should this be used for formal education or is it a personal/social tool better left in the realm of informal communication? From a practitioner's point of view, the challenge will come from the need to be flexible, adaptive, and innovative. In other words, the need is to rapidly develop new understandings of pedagogies to best utilize the person-to-person interactivity of emerging technologies (chapters 2 and 6). Some commentators go further and argue that new pedagogies are required. These pedagogies will respond in new, innovative, and pragmatic ways to disciplinary and contextual needs (chapter 5; Smyth, forthcoming).

In the past, videoconference appliances were technologically fairly similar. Although the brands differed, they basically transmitted and received audio, pictures of participants, and computer images. The move to Video over IP adds opportunities and complexity to this by

(a) permitting connections from webcams as well as from videoconference appliances,
(b) providing other functions such as collaboration through the use of digital canvasses, and
(c) providing applications that integrate video communications with telephony, text, computer applications, and social networking solutions using Presence.[3] These are often referred to as Unified Communications (UC).

Although videoconference has been used for some years, in many cases the use has not been informed by rigorous research leading to sound pedagogical practices. Videoconference has frequently copied typical lecture-hall formats of didactic information delivery rather than exploring approaches that are interactive and oriented towards knowledge construction. When considered alongside the combination of factors outlined above, the importance of social constructivist (Vygotsky, 1978) approaches to learning in higher education, which recommend such activities as peer collaboration, reciprocal teaching, and people learning from the experience of others (Schunk, 2000), should prompt much rethinking about the place of interaction in distance education. A key aspect of this is the consideration of approaches to capitalizing on the capacity of video communications to reduce isolation and increase the personalization of learning experiences for distance students. Indeed, there is now scope for the empowerment of distance learners and an opportunity to offer a much wider choice of strategies intended to enhance and support learning (Smyth, 2005; Smyth & Zanetis, 2007). Indications from the research literature are exciting. Many practitioners are beginning to explore the possibilities that video communications create, especially where connectivity is widely accessible by the vast majority of students.

Signal strength (or bandwidth) and picture quality will remain a challenge for those reliant on shared networks or satellite connectivity, such as in developing nations where internal mobile networks are proliferating rapidly on increasingly congested networks where bandwidth cannot be guaranteed. The personal experience of one of the authors in the Royal Kingdom of Bhutan indicates that connectivity is possible, but satellite up/download lag and poor signal strength result in ephemeral, small-sized images appearing without audio.

Further, in many cases, the management of videoconference installations has not been characterized by scalable and sustainable business models. Data from a current project investigating the use of video communications/rich media technologies across the university sector in Australia[4] confirms that many institutions are being left behind as video communications technologies proliferate outside the sector and are increasingly demanded by students and staff for use in learning and teaching. There are two aspects to this trend. First, institutions have generally acquired video communications technologies for particular projects or purposes and are just beginning to integrate them into institutional planning strategies, facilities, and teaching practice. Thus, cost-benefit analyses, business plans, and funding models are in their infancy and are often characterized by a lack of clear information about what is being used or planned for use across institutions. Second, policy support is similarly lagging alongside the lack of coherent operationalization and management strategies. Among many issues, readiness factors—including staff awareness of the potential for the technology as a pedagogical tool—are now becoming a focus for institutional planners.

Videoconference in Teaching and Learning

Videoconference has traditionally been seen as a tool for one-way transmission: lecturer to many students (Laurillard, 2002), although proponents of videoconferencing have long argued for its use on the basis of encouraging interactivity and interaction between participants (e.g., Andrews & Klease, 2002). This perception of transmission has been and remains a limiting factor for videoconferencing for teaching and learning activities even though there is a solid pedagogical basis for using it for guest lectures and other enhancement activities in distance learning. As the scope of videoconference grows beyond stand-alone rich media technology and into the realm of unified communications, preconceptions need to be shed in order to embrace the new capacity for engaging two-way video communication. Lecturers recognize the benefits of videoconferencing for a variety of purposes (Smyth, Stein, Shanahan, & Bossu, 2007), including higher degree research supervision, teaching to students on international campuses (Macadam, 2005), and research. Many are anxious for network connectivity and institutional infrastructures to enable seamless connections to remote students so that they can encourage student-to-student interaction. Internet connectivity has increased the potential for innovative pedagogy, signalling an opportunity for new rather than amended approaches (Smyth, forthcoming). Importantly, cost structures that bring connectivity within the reach of most distance students should further enable new pedagogies to emerge.

The place of video communications as tools for social constructivist approaches to teaching and learning is expanding as the reach of technologies extends via broadband, mobile, and wireless connectivity. As these tools increasingly extend to 3G phones and other mobile devices as outlined above, and the expectation for video and audio communication as part of the teaching and learning process increases, the demand for including mobile technologies in the learning process is growing.

This is creating the need for institutions to extend and strengthen wireless networks and to create learning and teaching spaces that are flexible enough to include the integration of these new technologies into teaching and learning activities.

Data[5] also show that videoconferencing is successfully used in situations where universities have established networked sites across multi-campus locations both nationally and internationally or for remote students to access university services from study centres. Videoconference is an important component of multi-location teaching in some institutions, such as the University of Wollongong, which has campuses on the south coast of New South Wales, and Central Queensland University, which has campuses in several central Queensland centres. Metropolitan institutions have been less enthusiastic adopters of video-conference despite the rapid uptake of learning management systems to enable off-campus study. However, there are some examples of metro-politan institutions using videoconferencing to teach to international campuses, either to provide a particular learning experience (e.g., in-digenous law at the University of Queensland) or to engage in twinning or other offshore teaching and learning agreements.

Other successful uses of video communications in distance education include the growing number of collaborations between universities and other institutions to offer courses or programs that traditionally have small numbers and have become commercially unviable for offer by a single institution.[6] These collaborations also offer opportunities for distance students to engage with a wider variety of experts and peers. The use of video communications in this way is leading to a merging of physical and virtual teaching and learning spaces. This will create further challenges for institutions in the design and fit-out of learning spaces.

Videoconference Technology

The cost structure of videoconference (and to a smaller degree, video communications) costs can be predicted with some degree of accuracy. However, before a meaningful discussion of these installations can begin, a conceptual understanding of the technology involved in videoconference is necessary. Videoconference appliances typically have:

> cameras to capture images of local participants,
> screens to display images of remote participants,
> microphones to capture local audio, and
> speakers to replay remote audio.

Videoconferences in their simplest form connect appliances at two locations. Such videoconferences are called *point-to-point* and may have one or more participants at each location, as shown in Figure 13.1. Often, however, more than two locations are connected to the same videoconference. In such cases, another device called a *bridge* or Multi-Conference Unit (MCU) is required to enable all participants to see and hear one another, as shown in Figure 13.2.

Figure 13.1
Point-to-point videoconferences

Figure 13.2
A multi-point videoconference

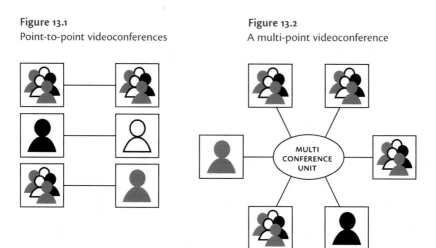

As Internet applications such as Skype continue to evolve, the infrastructure required to send and receive signals in point-to-point and multi-site conferences will become less significant, however, there will be a need for standards, or some bridging mechanism, if the increasing variety of commercial applications is to enable seamless connectivity between dedicated videoconference networks and Web-based applications.

Management of Videoconference

Sound business practices need to be involved in the management of video communications if they are to meet higher levels of use in a financially sustainable manner. Further, to cope with the predicted growth, video communications installations must be designed and installed in such a way to scale up to future usage levels without the need for expensive re-installations. Preliminary analysis of data gathered to date in the Leading Rich Media Implementation Collaboratively Project (2007–2009) funded by the Australian Learning and Teaching Council (ALTC) indicates a trend towards the centralization of management into information communications technology directorates, and concerns with the lack of information about the proliferation of brands, forms of UC technology, and realistic plans for ongoing issues, such as the cost of maintenance contracts and the use of recurrent funding for technology purchases.

Cost and scalability

To effectively manage an expanding videoconference installation, a clear idea of costs is essential. The costs involved with owning a videoconference installation fall into several areas: videoconference equipment purchases, network or traffic costs, training, operation, and maintenance. Financial planning for videoconference installations can be reasonably straightforward as the costs of endpoints and Multiple Conference Units (MCUs) are relatively predictable.

Apart from deals that may be done — for example, for the supply of multiple units — videoconference endpoints have discrete costs that vary from brand to brand. So it is quite simple to budget for endpoints in an expanding videoconference system: the more endpoints the higher the cost. This is illustrated in Figure 13.3 as the cost of each additional endpoint is represented as a step in the graph. Scalability of MCUs is only slightly more complex than that of endpoints. MCUs generally have a fixed number of ports, which indicate the number of concurrent connections. That is, a twenty-port MCU can connect up to twenty endpoints in a number of discrete conferences. It can bridge one conference of up to twenty endpoints, four conferences of five endpoints, or multiples of conferences and endpoints that are equal to or less than twenty. The first step in Figure 13.4 indicates the cost to purchase a MCU as a fixed amount for three connections up to twenty (for example). After that, another MCU is needed, and hence the costs step up once more. For applications with only a small number of connected endpoints (often less than five), endpoints can be supplied with a multi-point function built in.

Figure 13.3 Scalability of endpoint costs   **Figure 13.4** Scalability of MCU capacity

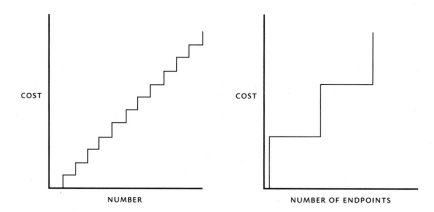

While equipment costs can be easily predicted and planned, as shown, the costs of support can vary widely and depend on many factors. These factors include the operational skills of videoconference participants and the size of the installation. These factors do not operate in isolation, and often there can be efficiencies gained by training participants in basic operation skills and then adopting a centralized approach to support. Distributed support is defined as the presence of a technical or operational support person and replacement equipment at each endpoint. This is in contrast to centralized support, where the technical/operational support person or persons are in one location and communicate with the other locations via telephone, videoconference, or other means. Centralized or distributed support personnel can access the MCU and hence assist in all aspects of the videoconference.

A centralized support function usually requires the purchase of an additional endpoint and computer equipment to access the MCU. Thus, centralized support has a fixed associated cost. For small installations, the distributed support model can be cheaper than a centralized one, as the support person uses the local participants' endpoint. However, there is a point where distributed support becomes more expensive than centralized support. This is shown in Figure 13.5 where the two lines intersect.

Figure 13.5 Support costs

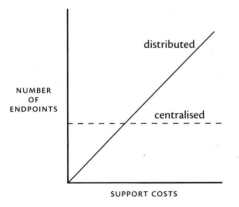

In distance education, videoconferences are often used repeatedly during a teaching semester. In such cases where the participants will be using the videoconference technology frequently, cost efficiencies can be attained by training participants in the operational aspects of the videoconference technology, as this will reduce the need for operational support. This approach has pedagogical benefits as well, as when students are confident with the technology, they are likely to use it more effectively. For example, newcomers to videoconference are often reluctant to use the controls to frame the image they are sending. This can result in small images of participants on a large screen of little or no communicative value. However, when students are comfortable with the technology, they are more likely to frame reasonable images of themselves, thus optimizing the communicative value of the videoconference.

Sustainability considerations

At first, videoconference was marketed to organizations as a way to reduce costs through the reduction or elimination of travel. In recent times this has been complemented by a marketing approach that touts videoconference as a means to reduce organizations' carbon footprints, and some videoconference manufacturers use this as a central sales point. For example, see the Green Manifesto published by Tandberg, a manufacturer of videoconference technology (Tandberg, 2008). For distance education or multi-campus institutions, the cost-benefit analysis is a relatively simple one in which the costs of travel to regional centres or the employment of academic staff at those centres is compared to the costs of videoconference equipment purchases, training, operation, and maintenance. However, little reliable data are available to aid decision-making. There is a need for work to be done to develop and disseminate viable and sustainable models for adopting video communications.

Future Research

It appears safe to assume that the use and uptake of videoconference technologies will grow in the near future. A further safe assumption is that other video communications tools will become mainstream mechanisms of communication in organizations. These tools include Unified Communications and in particular Collaboration. UC was referred to earlier as the bringing together of a range of existing and new communications applications, and the hype suggests that businesses can benefit from them. One vendor suggests that UC "combines all forms of business communications into a single, unified system that provides powerful new ways to collaborate" (Cisco, 2008). Another states that Unified Communications "will transform business in the coming decade" (Microsoft, 2008).

If UC has transformative advantages for business, the question must be asked: can UC deliver similar advantages for distance education? The recent Wainhouse Research Segment Report (Greenberg, 2008) indicates that "these technologies [are] uniquely suited to distance and e-learning" (p. 11). However, the following questions must be asked:

> Are these aspects of UC appropriate for distance education?
> What benefits do these aspects of UC bring to distance education?

Unified Communications typically include video communications that use webcams and computers, thus facilitating video communications from participants' desktops, and it is easy to imagine the efficiencies that are possible with this technology. The communications can be unified with collaboration tools that use digital canvasses, also on desktops. In the context of distance education, these questions must be asked:

> Will UC replace appliance-based videoconference?
> What is the role of online collaboration (with video communications)?

Perhaps one of the greatest technological challenges will be provided by the single-user model of UC; that is, one person to a computer or endpoint. While this is not a problem for small meetings of up to eight to ten participants (and hence eight to ten windows on each participant's screen), how will it work or indeed can it work in distance education? Preferably, the single-user model should become a tool for group work and few-to-few communications that enhance learning. The benefit of visual interaction is lost when participants cannot clearly see others' faces, so caution is encouraged when considering uses of such applications.

Telepresence is yet another video communications technology that has not been fully explored for use in distance education. Telepresence installations are very expensive to purchase and operate, and generally cater to meetings of twelve or fewer participants. The experience of meeting in telepresence is very close to that of meeting face to face, as the pictures are of high quality. Further attention to furnishings and room layout creates images and sounds of the distant participants that are lifelike. In business, telepresence installations are effectively used for small meetings. As they cater to small numbers, the use of telepresence installations in distance education is probably limited and the cost benefit not as positive as that of videoconference. However, their use in distance education should not be discounted without thorough investigation.

While there has been some limited investigation of the role of mobile technologies in distance learning, this is an area that is set to expand rapidly, particularly in relation to video communications. The expansion of wireless networks and the increasing flexibility and power of mobile devices means that these mobile technologies will become common in the distance education landscape. However, there are many challenges for institutions in adopting these technologies for teaching and learning, including managing potentially "disruptive" technologies and finding ways around the current lack of interoperability between mobile devices.

Clearly, significant levels of investigation are required if, in the future of video communications in distance education, these new communications technologies and their integration with computer applications are to be recognized and taken advantage of.

Conclusion

The use of video communications for distance education is experiencing a revival — perhaps a renaissance of sorts — as the growing availability and reliability of such technologies and improvements in cost structures enable the widespread use of the Internet to support such technologies. Mobile technologies, student expectations, and the increasing flexibility of teaching and learning activities within and between institutions is further driving the uptake of video communications for distance and other forms of teaching. However, to ensure the sustainability and scalability of the use of video communications, many factors need to be given thoughtful consideration, in particular the development of business models to thus support these technologies. The development and implementation of appropriate pedagogies that utilize the interactive and collaborative capabilities of video communications technologies is a vital aspect of the successful use and sustainability of these tools.

NOTES

1 ISDN is the Integrated Services Digital Network, a network of digital communications lines operated by telecommunications companies.

2 Data from UNE network administrator

3 Presence is an application in Unified Communications suites (and also found in many video communications applications such as Skype) that reports the availability status of individual users to the rest of the networked community. Typical Presence options are: available, away, and busy, but when a participant's presence indicator is linked to their diary the Presence can indicate when the participant is expected to be available.

4 Leading Rich Media Implementation Collaboratively: Mobilising International, National and Business Expertise funded by the Australian Learning and Teaching Council (survey data from 19/37 universities).

5 Ibid.

6 Ibid.

REFERENCES

Andrews, T., & Klease, G. (2002) Extending learning opportunities through a virtual faculty: The videoconference option. *International Journal of Educational Technology, 3*(1). Retrieved 1 December 2008, from http://www.ascilite.org.au/ajet/ijet/v3n1/andrews/

Caladine, R. (2008a). *Enhancing E-Learning with Media-Rich Content and Interactions*. Hershey, PA: InfoScience.

Caladine, R. (2008b). *An Evaluation of the Use of Peer-to-Peer Real-Time Communications Applications within the Australian Academic and Research Community*. Report posted on AARNet. Retrieved 1 December 2008, from http://www.aarnet.edu.au/Blog/archive/2008/07/03/peer-to-peer_RTC.aspx

Cisco. (2008). *Unified Communications*. Retrieved 2 December 2008, from http://www.cisco.com/en/US/netsol/ns151/networking_solutions_unified_communications_home.html

Greenberg, A. (2008). *The Distance Education and e-Learning Landscape. Volume 2: Videoconferencing, Streaming and Capture Systems for Learning* (Wainhouse Research Segment Report). Brookline, MA: Wainhouse Research.

Laurillard, D. (2002). *Rethinking University Teaching: A Framework for the Effective Use of Educational Technology* (2nd ed.). London: Routledge.

Macadam, R. (2005). *External Evaluation Report on the First EU/Australia Pilot Co-operation in Higher Education Project*. Sydney, Australia: Learning through Exchange — Agriculture, Food Systems and Environment (LEAFSE).

Microsoft. (2008). *Microsoft Unified Communications Vision*. Retrieved 2 December 2008, from http://www.microsoft.com/uc/vision.mspx

Nipper, S. (1989). Third-Generation Distance Learning and Computer Conferencing. In Mason, R. & Kaye, A. (Eds.), *Mindweave: Communication, Computers, and Distance Education*. Oxford, UK: Pergamon.

Schunk, D.H. (2000). *Learning Theories: An Educational Perspective* (3rd ed.). New Jersey: Merrill.

Smyth, R. (2005). Broadband videoconferencing as a tool for learner-centred distance learning in higher education. *British Journal of Educational Technology, 5.*

Smyth, R. (forthcoming). Theorising an interactive model for technology enhanced learning using rich-media in higher education. *British Journal of Educational Technology.*

Smyth, R., Stein, S.J., Shanahan, P., & Bossu, C. (4–6 September 2007). *Lecturers' Perceptions of Videoconferencing as a Tool for Distance Learning in Higher Education*. Paper presented at the ALT-C 2007: Beyond Control: Learning technology for the social network generation, Association for Learning Technology 14th International Conference, Nottingham, UK.

Smyth, R., & Zanetis, J. (2007). Internet-based videoconferencing for teaching and learning: A Cinderella story. *Distance Learning, 4*(2), 61–70.

Tandberg. (2008). *A Manifesto for Corporate Action*. Retrieved 1 December 2008, from http://www.seegreennow.com/TandbergManifesto.aspx

Taylor, J. (2001). *Fifth-Generation Distance Education*. Presentation to the Australia-European Union Roundtable on Cooperation in Education and Training, Canberra.

Vygotsky, L. (1978). *Mind in Society*. London: Harvard University Press.

Part 4
Learner-Learner, Learner-Content, and Learner-Instructor Interaction and Communication with Emerging Technologies

USING SOCIAL MEDIA TO CREATE A PLACE THAT SUPPORTS COMMUNICATION

14

> *Rita Kop*

Abstract

The rapid development of the Internet and its emerging peer-to-peer technologies make different structures and educational organizations and settings a possibility. Proponents of Web 2.0 technology suggest that these tools could facilitate the transformation from an education model that is structured and controlled by the institution using a "broadcasting" model in an enclosed environment, to a model that is adaptive to learners' needs. This chapter will analyze the changing nature of communication in distance education through the lens of an online learning program that created a comfortable online "place" by using Web 2.0 tools extensively. The aim of the research was to establish if the goal to build a learning and teaching model where learners increasingly take control and share information would be a realistic one.

Introduction

Much discussion has taken place about the impact of the introduction of virtual learning systems on education. The changed position of educational institutions such as universities due to the changed sense of space, place, and identity in a virtual learning environment has been lamented as a loss, as universities were seen as places where people came together, where minds met, and where new ideas were conceived as nowhere else in society (Edwards & Usher, 2000). Some academics have expressed reservations about the networked alternatives

(Greener & Perriton, 2005), while proponents of the use of Web 2.0 technology in education have argued that tools such as wikis and blogs could facilitate this networking role, as the openness of the media and the willingness of people to share in such experiences encourage a similar discussion of ideas and collaborative development of thoughts and knowledge. The added advantage of the online tools would lie in their globally positioned communication forums, which would provide immediate responses on a scale unimaginable in the traditional university (Siemens, 2008).

The chapter will discuss the use of Web 2.0 technologies — blogs as reflective journals, wikis for group activities, chat to enhance the affective and social aspects of learning, and pod and video casts by students and tutors — to create a presence and immediacy of interaction. A tutor in this context is the person who is teaching the course, but who has also written the course materials and who has worked with the learning technologist to choose the most appropriate technologies and media to convey the subject knowledge. The chapter will explore the roles of the tutor, learner, and learning technologist in an online learning environment that was initially controlled by the tutor and the institution, and is increasingly adapted to become a place where the learner can feel confident to learn autonomously. In addition, the chapter will highlight the challenges and solutions a university department of adult continuing education faces in supporting students from socially, economically, and geographically disadvantaged backgrounds by using Web 2.0 tools in learning and teaching in a distance education environment.

Communication in Online Learning

Since antiquity, communication and dialogue have been seen as the crucial components in the creation of knowledge. Communication technologies, however, seem to be changing their nature. Dewey (1958) identified communication as the most important aspect in making people what they are: "mind, consciousness, thinking, subjectivity, meaning, intelligence, language, rationality, logic, inference, and truth — all of these things that philosophers over the centuries have considered to be part of the natural 'make-up' of human beings — only come into existence through and as result of communication." By communication with others, our inner thoughts will become clear: "It is because people share in a common activity, that their ideas and emotions are transformed as a result of and in function of the activity in which they participate" (Biesta, 2006, pp. 17–19). Online communication is quite different from face-to-face communication: online messages may be one-way, the receiver might not know the sender, his or her intentions, or if s/he can be trusted (Kop, 2006). Online communication is a fast connection between systems and networks, conveying messages produced by people. The mediated nature of online communication has been seen as problematic by many practitioners, particularly while using discussion boards in Virtual Learning Environments (VLEs). Issues of power and control, lack of autonomy, high level of tutor support required, and lack of options for personalization, on top of affective issues, have been seen to influence learning in a negative way (Kop, 2006; Mann, 2005; Mason & Weller, 2001). Conrad (2007) noted that some students may be overwhelmed by the number of messages on the discussion board, while other students may experience monologue confessions and tensions in online group activities. Salmon (2004) and Gulati (2006) emphasized the need for quality tutor engagement on discussion boards to cultivate a non-threatening and supportive community. Moreover, as Mayes suggests, "activity, motivation, and learning are all related to a need for a positive sense of identity shaped by social forces," which is hard to achieve by using VLEs (Mayes, 2002, p. 169). A good VLE brings information and communication together and offers some structure to the tutor and the "not so technologically adept" learner. Good communication takes place through tasks set by the tutor, and because learners are dispersed, this will inhibit the forming of trust relations in the learning space and affect the quality of the communication and subsequently of the nature of the learning (Kop & Woodward, 2006).

The relationship the learner has to the community in which he or she learns is a determining factor in the learning process (Dron & Anderson, 2007); the more active the engagement in group communication the better. Emerging Web 2.0 technologies facilitate this development as they create an immediacy that has been missing from the VLE (Gur & Wiley, 2007). Podcasts and videocasts could be used to directly speak to tutor and learner, participants in the learning experience, and foster a close connection. Chat has also been highlighted as a powerful tool to create presence and to enhance the social and affective engagement (Carroll, Kop, & Woodward, 2008). Even though chat might not be seen as a Web 2.0 tool, its use in a social context and in combination with other Web 2.0 tools makes it into a powerful tool to enhance the learning experience. The idea of "transactional nearness" (the closeness, not in person per se, but in exchange of ideas by participants in a learning experience) resonates with the thoughts of critical educators such as Freire and Macedo (1999, p. 48) who emphasized that tutors should have a directive role. In this capacity, tutors would enter into a dialogue "as a process of learning and knowing" with learners, rather than the dialogue being a "conversation" that would remain at the level of "the individual's lived experience." Freire (1999) felt that this capacity for critical engagement is not present if educators are reduced to facilitators.

Siemens (2008) argues that the distributive effect of communication is an important feature of the new wave of emerging technologies. He highlights that networks offer opportunities for learning. He does not make clear, however, how the transactional distance between people on networks would affect learning. Dron and Anderson (2007) gave the transactional distance between participants considerable thought and make a clear distinction between learning in groups, networks, and collectives. They argue that there is a difference in "presence" and subsequent engagement in these three entities. They see the level of emotional engagement and presence being the highest in groups, which would be the typical classroom or distance education student group. Engagement and presence would be lower in online networks (e.g., the blogosphere or large, informal, online courses such as Connectivism and Connective Knowledge 2008/2009 [Siemens & Downes, 2008]); and the lowest in collectives, such as on sharing sites that use tags as connection, such as Flickr for photos, or Delicious for bookmarks.

Dron and Anderson (2007) emphasize that there is an interrelationship between "engagement in learning," "transactional nearness," and "emotional involvement": the closer the feeling of connectedness, the higher the level of commitment to the learning activity; the closer the relationship with the people involved in the undertaking, the higher the inclination to engage in communication and learning.

Recent research in distance education has shown that a high level of interactional experience (i.e., dialogue between teacher and learner and among learners) is what leads to a rich and engaging online learning experience (Carroll et al., 2008). Moreover, Gur and Wiley highlight (2007, p. 7) that "instructional designers need to create structures in which a caring relationship might be enhanced and a dialogue can take place." This is exactly what Carroll et al. (2008) argue for when they call for the development of a "Learning Place" as opposed to a VLE in online learning. They discuss Oldenburg's concept of the Third Place (Oldenburg, 1989), and the ideas of Fisher, Durrance, and Hinton (2004) of an Information Ground. What these physical places have in common is that they are informal and that people feel comfortable being there. People come together and interact in informal places such as doctors' offices or cafés, and Carroll et al. (2008, p. 3) highlight that "these environments could be transferred to the online space and are well suited to informal online learning, where learning occurs by chance and where participants feel comfortable in building their own presence while communicating with others." In fact, these spaces have now materialized online in the form of online social networking tools, including YouTube, MySpace, Facebook, blogs, and wikis — tools that might be implemented and used in distance education.

Web 2.0 Technologies in Distance Education

Some argue that emerging technologies, and Web 2.0 tools in particular, with their intrinsic participatory features and user control, have changed the e-learning landscape (Duffy 2008) and are able to facilitate a new "pedagogical paradigm." Others, however, question the level of "higher-order thinking" that can be achieved through these tools, as communication — if not directed by a critical tutor — might remain at a superficial level (Kop & Hill, 2008).

A number of academics have shown an interest in blogs and have seen their potential in an educational environment (Halavais in Glaser, 2004; Lankshear & Knobel, 2006). Other academics and researchers have noted the bias and unreliability of material written in blogs, wikis, and personal spaces, but value the lack of restrictions and control by institutions so ideas can be freely expressed through the tools (Downes, 2003).

Educators have embraced blogs and wikis as tools for debate and have found out that they work differently from the traditional academic environment. Walker (in Glaser, 2004) notes:

> Blogging alongside other academics in my field ... is a form of indirect collaboration.... . There is an openness and a willingness to share in blogging ... that means I know more about many of my fellow bloggers' research than I do about a colleague whose office is down the corridor. (p. 1)

Other comments from lecturers and tutors about the use of blogs in their classes include: "the push into critical thinking, critical reading, and reflection" (McIntire-Strasburg, 2004); "the ability to achieve active back-and-forth discussions outside the classroom" (Martin & Taylor, 2004); "Students are blogging about topics that are important to them. Students direct their own learning while receiving input and feedback from others" (Ferdig & Trammell, 2004). Mason (2006) used blogs for peer commenting and saw a high level of reflection as one of the positive effects of blogging as part of a course. On the negative side, Mason also saw blogs' potential for shallowness and meaninglessness.

Lamb (2004) notes the openness of the wiki environment. He sees a number of possibilities to use wikis in an educational context: as spaces for brainstorming; as collaborative areas for teams to work on projects, outlining and managing activities or research; and as repositories of shared knowledge. Additionally, James (2004) and Lamb (2004) indicated the need for teachers to hand over control over content in using wikis to ensure successful knowledge-building. The role of the tutor would lie in "setting the scene" and thinking up problems relating to the subject being taught, while allowing students to develop the wiki to their own liking.

While video podcasts seem to have been mainly used to explain difficult concepts, not much research could be found on their use in educational contexts. Kamel Boulos and Wheeler (2007) highlight the possibilities of podcasts in the creation of scenarios in health-care education, while Savel, Goldstein, Perencevich, and Angood (2006) report on the use of podcasts for the fast and cheap delivery of media content. It is through access to broadband, the availability of easy-to-use and freely available audio and video production and distribution software, and the explosion of social video sites, such as YouTube, that students and tutors can more easily become creators and distributors of video content outside the institutional control.

Jenkins (2007) argues that academia will have to engage with these technologies that are widely used outside educational institutions. He states that "the best thinking (whether evaluated in terms of process or outcome) is likely to take place outside academic institutions — through the informal social organizations that are emerging on the Web" (p. 1) and he would like educators to get involved in the discussion by using the tools. The challenge, of course, is to use the tools in a meaningful way that will enhance education.

The Research

In-depth research in Swansea, South Wales, UK, in the use of Web 2.0 technologies in an online adult education program had a particular focus on using Web 2.0 technologies' "true dialogue" potential. The research conducted was a case study investigation in a project that developed and taught an online Higher Education Certificate at the undergraduate level, mainly at a distance, and made a concerted effort at creating an online "place" in which students would feel comfortable using Web 2.0 tools. The research results presented in this chapter will draw on semi-structured interviews with three of the tutors, three learning technologists, and nine students at different stages of the two-year-long program, together with an analysis of the activities and interactions on the learning environment over a fifteen-month period. Particular attention was paid to the use of blogs, (video) podcasts, wikis, and chats. Interviews and online environment interactions were recorded, transcribed, coded, and analyzed using standard content analysis techniques (Hammersley et al., 2001). The students were non-traditional adult learners, employed (quite often under-employed) in small and medium enterprises and social enterprises dispersed over the South West Wales and the Valleys region of the UK.

The project took place in a "brick and mortar" university, and a curriculum was created relevant to student needs: in addition to business studies modules, course topics ranged from creativity, reflection, and action-research to critical thinking and information literacy. The tutors, together with the project team, decided on the course content, worked directly with learning technologists, and were also the ones teaching the students. The first module was classroom-based. It was a skills-based course in which learners were made familiar with communication tools such as chat and discussion boards and the application of Web 2.0 tools such as blogs, wikis, and the production of podcasts and videocasts. This was done to ensure a high level of proficiency in using the tools, as it was envisaged that throughout the program these tools would be used for communication and for the submission of assignments. A process of negotiation between tutors, learning technologists, and students was at the heart of the development, as learning technologists and the project manager were the tutors on this first module, while the technologists would provide help-desk support to students during the online modules to ensure that a level of trust was developed from the very beginning. The person responsible for the design of the (Moodle) leaning environment was an expert in Human Computer Interaction, as it was felt that the technology should take into consideration the overall impression, feelings, and interactions that a user has; it is about supporting the creation of relationships with individuals and creating an environment that connects on an emotional level in addition to providing enough opportunities for both the learner and the tutor to create an exciting and dynamic learning experience. Therefore, from the start, the project team has been very aware of the need to design a learning space where people would feel comfortable and in which all participants could have a "presence" (Carroll et al., 2008).

In the online place, several communication tools were used to create a sense of presence and for social engagement. These were "the lounge" (an informal place in which participants could chat, watch a course-related video, or give each other links to materials and information), discussion boards, chat spaces, wikis, podcasts, and blogs in the form of reflective diaries. It was quite often the combination of tools used that made for a quality learning experience. Clearly some tools are better than others at enabling the provision of individual feedback, while others are better at facilitating group work, or to ensure the social interaction of a group of students. In this program, blogs were used as reflective diaries, which were used extensively. Students were very open and honest in what they wrote about their learning experience. The tutors used the comment feature to give students personal feedback. Some tutors were particularly good at providing feedback, and on one particular course, it was clear that the students' confidence levels and their knowledge and eagerness to participate increased because of the personal approach to feedback in the diaries. In the words of one of the students, "independently, you don't know whether you're making any sense or not. Then the tutor comes back and says: 'Yes, you've got the point' and will give you another kind of perspective on it then. Then you feel like you're taking a step forward" (Student 9). The journal entries show that the students were benefiting from the strong tutor presence and were building up their understanding of the material and subject area, as expressed by student 3: "Tutor 1: Thanks for the recent feedback. I'm really heartened by your comments … I'd like to say that I really enjoyed this module, your openness, and positive support. Thanks/Diolch!"

It was through reading the personal experiences of students in the diaries, rather than the collective discussions through other tools, that tutors decided to produce videocasts and include these on the learning place. They used two types of videos: the first one to clarify concepts, and the second to support students — to raise the level of confidence or lessen anxiety at particular moments in the course. Tutors required a high level of reflection on the learning and teaching taking place in the course, in addition to being willing to show themselves as real human beings, rather than as distantly removed tutors, as the videos were very personal accounts and observations about the course progress. Videos of this nature, made on the spur of the moment, were very much appreciated by students, who said the immediacy that the videos created made them feel they were part of a group of people they felt close to. This also led to students themselves responding with videos in different situations, as part of discussions or to reveal more about themselves.

> "The use of video has had a positive effect on building up the pro-gramme's online place; it has offered a multi-sensory approach to knowledge-sharing, reflection, and communication of ideas, which in turn has enhanced the relationship between tutors and students" (Carroll et al., 2008, p. 156). At that point in the program, some of the students were gaining confidence and were questioning the use of Web 2.0 tools: "The latest wiki in Section 2 of Action Research has unfortunately frustrated me, as once again, separate entries have been made" (Ibid., p. 157).

Wikis are seen to be valuable in carrying out collaborative activities involving the creation of a joint piece of work, but in the program this rarely worked well. The adult learners weren't used to the concept of "collaborative knowledge creation," and in most cases preferred to see visibly what their contribution to the learning task had been, rather than for it to become a joint venture. This resulted in discussion-board style participation. The idea of making changes to contributions by others was alien to students and they did not feel comfortable doing this. Lamb (2004) and James (2004) argue that for wikis to work well, the control should be handed over to the learners, but even though most of the students on the course were young adults, they didn't feel at ease at all with shaping documents collaboratively, in particular in tasks involving developing concepts, rather than more practical tasks. It was not until the tutor became involved and showed that it would be fine to make changes to the document by doing so herself, while also reinforcing the ground rules about using wikis, that students would engage collaboratively in producing a document.

Another major problem in the use of wikis has been their asynchronous nature, in addition to the different time management of students and their level of commitment to contributing at times that suited others in order to finalize a particular task on time. Not all activities carried out through wikis had positive outcomes. Conceptual thinking in a collaborative fashion was clearly much harder than the use of wikis to organize events.

Extensive use of weekly chats has been made in all modules, especially to create a sense of "togetherness" and to facilitate social interaction. After using a variety of Web 2.0 tools and more traditional ones such as discussion boards, all tutors involved mentioned that they had a good feel for who the students were as persons. The use of the chat tool and the reflective diaries in particular helped to foster affective relations.

This research indicates that before adult learners feel confident enough to venture to engage in online networks in order to find information and to communicate, they first need nurturing by a tutor who is genuinely interested in them as persons and in their learning. Yet, there is a fine balance between supporting and "letting go." As Bouchard (2009) emphasizes, there are different aspects in and levels of self-direction while learning on semi-autonomous learning systems that adult students will have to reach before they feel comfortable directing their own learning.

Conclusion

Information Communication Technology is swiftly changing; new applications and innovations come to the fore nearly every day. The way in which global networks and communities of interest are currently being formed through emerging technologies is encouraging people in developing new and different forms of communication outside formal education.

This research has shown that communication and interaction with other learners and with tutors is at the heart of a quality online learning experience, and that this mixed use of multi-sensory Web 2.0 tools in a flexible manner, "if and when required," can be very powerful. By creating an online place where people feel comfortable and relaxed, a place that affords communication and interaction at different levels and while using a variety of tools, both tutors and students develop a strong sense of presence that can help participants gain confidence in both their learning and teaching. The direct interaction between the designers and writers of content and the learners in an environment that allows for affective involvement and "transactional nearness" ensures that meaningful activities are created and meaningful communication can take place. The emergence of Web 2.0 tools and their combined use can help in the facilitation of authentic interaction and communication in the learning process.

The future of online learning lies in the hands of the empowered tutor, who has the control to send out podcasts to students or set up a wiki when s/he feels this will help in their learning; the engaged student, who also has the power to bring resources s/he finds on the Internet into the learning arena or to produce a video or sound file to make his/her voice heard in the learning community; and the negotiating learning technologist who through his/her close involvement with the learning process can help to facilitate technical needs. Through a flexible approach, where some control and standardization is being relinquished by the institution, the educational establishment will begin to allow new technological tools to be used to their full capacity (chapters 10, 11). More research is required into the dynamics of the three actors in their development and experience of online learning, as the level of control imposed by the institution in an era where the affordances of Internet tools exert pressure for a different distribution of power in the educational arena.

Two other issues would be worthwhile exploring. Firstly, an examination of the impact of the changing communications environment on the quality of the learning experience and quality of knowledge created; this in particular in relation to the depth of communication achievable through the use of social software. Secondly, an investigation into the level of control imposed by the institution on the learner, in addition to aspects of learner autonomy that would be desirable in order for students to thrive in the evolving and participatory model of distance education described in this chapter and in chapters 3, 6, and 7. Emerging technologies enable the transfer of more and more teaching tasks to the learner; yet, the learner may not necessarily be ready to accept these as his or her own without the help of a tutor.

REFERENCES

Biesta, G. (2006). Context and Interaction: Pragmatism's Contribution to Understanding Learning-In-Context. Seminar 4, ESRC Teaching and Learning Research Programme, Thematic Seminar Series, University of Stirling, Edinburgh.

Bouchard, P. (2009). Some factors to consider when designing semi-autonomous learning environments. *Electronic Journal of e-learning, 7*(2), June 2009, 93–100.

Carroll, F., Kop, R., & Woodward, C. (2008, November). Sowing the seeds of learner autonomy: Transforming the VLE into a Third Place through the use of Web 2.0 tools. In *ECEL-European Conference on e-Learning* (pg 152–159), University of Cyprus, Cyprus.

Conrad, D. (2007). The pedagogy of confession: Investigating the online self. In Osborne, M. et al. (Eds.), *The Pedagogy of Lifelong Learning: Understanding Effective Teaching and Learning in Diverse ontexts* (pg 203–213). London, UK, and New York, NY: Routledge.

Downes, S. (2003). Weblogs at Harvard Law. *The Technology Source*, July/August 2003, Retrieved 23 October 2009, from http://technologysource.org/article/weblogs_at_harvard_law/

Dron, J., & Anderson, T. (2007). Collectives, networks, and groups in social software for e-learning. *World Conference on E-Learning in Corporate, Government, Healthcare, and Higher Education (E-LEARN) 2007*, Quebec City, Quebec, Canada.

Duffy, P. (2008). Engaging the YouTube Google-eyed generation: Strategies for using Web 2.0 in teaching and learning, *Electronic Journal of E-Learning, 6*(2).

Edwards, R. & Usher, R. (2000). *Globalisation and Pedagogy: Space, Place, and Identity*, London and New York, NY: Routledge.

Ferdig, R. & Trammell, K. (2004, February). Content delivery in the "blogosphere." *Technological Horizons in Education Journal* (pp. 1–7). Retrieved January 2005, from www.thejournal.com

Fisher, K., Durrance, J.C., & Hinton, M.B. (2004). Information grounds and the use of need-based services by immigrants in Queens, NY: A context-based, outcome evaluation approach. *Journal of the American Society for Information Science & Technology*, 55(8), 754–766.

Freire, P., & Macedo, D. (1999). Pedagogy, culture, language, and race: A dialogue. In Leach, J. & Moon, B. (Eds.), *Learners and Pedagogy*. London: PCP.

Glaser, M. (2004). Scholars discover weblogs pass test as mode of communication. Posted to *Online Journalism Review*, USC Annanberg, 11 May 2004, Retrieved 28 September 2006, from http://ojr.org/ojr/glaser/1084325287.php

Greener, I., & Perriton, L. (2005). The political economy of networked learning communities in higher education. *Studies in Higher Education*, 30(1), pp. 67–79.

Gulati, S. (2006). Learning during online and blended courses [Doctoral thesis]. City University, London. Retrieved from http://www.yourlearning.com/phdresearch.html

Gur, B., & Wiley, D. (2007). Instructional technology and objectification. *Canadian Journal of Learning and Technology*, 33(3). Retrieved 17 October 2009, from http://www.cjlt.ca/index.php/cjlt/issue/view/20

Hammersley, M., Gomm, R., Woods, P., Faulkner, D., Swan, J., Baker, S., Bird, M., Carty, J., Mercer, N., & Perrott, M. (2001). *Research Methods in Education Handbook*. Milton Keynes, UK: Open University Press.

James, H. (2004). My Brilliant Failure: Wikis In Classrooms. Message posted to Kairosnews, weblog for discussing rhetoric, technology, and pedagogy. Retrieved January 2005, from http://kairosnews.org/node/3794

Jenkins, H. (2007). From YouTube to YouNiversity: The Official Weblog of Henry Jenkins. Message posted to http://www.henryjenkins.org/2007/02/from_youtube_to_youniversity.html

Kamel Boulos, M.N., & Wheeler, S. (2007). The emerging Web 2.0 social software: An enabling suite of sociable technologies in health and health care education. *Health Information & Libraries Journal*, 24(1), 2–23.

Kop, R. (2006, September). Blogs, wikis, and VOIP in the virtual learning space: The changing landscape of communication in online learning. In Whitelock, D. & Wheeler, S. (Eds.), *ALT-C 2006: The Next Generation Research Proceedings, 13th International Conference* (pp. 50–56). Edinburgh.

Kop. R., & Hill, A. (2008). Connectivism: Learning Theory of the Future or Vestige of the Past? In *The International Review of Research in Open and Distance Learning*, 9(3).

Lamb, B. (2004). Wide open spaces: Wikis, ready or not. *EDUCAUSE Review*, 39(5), 36–48.

Lankshear, C., & Knobel, M. (2006, April). Blogging as participation: The active sociality of a new literacy. *American Educational Research Association Conference*, San Francisco, CA.

McIntire-Strasburg, J. (Fall 2004). Blogging back to basics. *Lore: An E-Journal for Teachers of Writing*. Retrieved 2 March 2007, from http://www.bedfordst-martins.com/lore/digressions/index.htm

Mann, S.J. (2005). Alienation in the learning environment: A failure of community? *Studies in Higher Education, 30*(1), 43–55.

Martin, C., & Taylor, L. (Fall 2004). Practicing what we teach: Collaborative writing and teaching teachers to blog. *Lore: An E-Journal for Teachers of Writing*. Retrieved 2 March 2007, from http://www.bedfordstmartins.com/lore/digressions/index.htm

Mason, R. (2006, July), Learning technologies for adult continuing education. *Studies in Continuing Education, 28*(2), 121–133.

Mason, R., & Weller, M. (2001). Factors affecting students' satisfaction on a Web course. *Education at a Distance, 15*(8).

Mayes, T. (2002). The technology of learning in a social world. In R. Harrison, R. Reeve, A. Hanson, & J. Clarke (Eds.), *Supporting Lifelong Learning: Perspectives on Learning* (pp. 163–175), London: Routledge Falmer.

Oldenburg, R. (1989). *The Great Good Place*. New York, NY: Paragon Books

Salmon, G. (2004). *E-Moderating: The Key to Teaching and Learning Online*. London: Routledge Falmer.

Savel, R.H., Goldstein, E.B., Perencevich, E.N., & Angood, P.B. (2007). The iCritical Care Podcast: A novel medium for critical care communication and education. *Journal for the American Medical Informatics Association, 14*(1), 94–99.

Siemens, G. (2008). Learning and knowing in networks: Changing roles for educators and designers. Paper 105: University of Georgia IT Forum. Retrieved 17 January 2009, from http://it.coe.uga.edu/itforum/Paper105/Siemens.pdf

Siemens, G., & Downes, S. (2008). Connectivism and Connected Knowledge course. Retrieved 27 November 2008, from http://ltc.umanitoba.ca/wiki/Connectivism

15 TECHNICAL, PEDAGOGICAL, AND CULTURAL CONSIDERATIONS FOR LANGUAGE LEARNING IN MUVEs

> *Charles Xiaoxue Wang, Brendan Calandra, & Youngjoo Yi*

Abstract

This chapter explores various facets of language learning within Multi-User Virtual Environments (MUVEs). As context, we present a recent endeavour to investigate the use of the MUVE *Second Life* to connect two groups of university students in the United States and China. We provide theoretical supports, the context of our work, and some considerations for creating and facilitating similar environments.

Introduction

Scholars and practitioners alike believe that technological advancement together with fast-growing international ventures, increasing business outsourcing, and expanding distance education opportunities have dramatically changed training and learning landscapes (Rosenburg, 2001; Welsh, Wanberg, Brown, & Simmering, 2003). Successful technologies afford us the ability to accomplish certain tasks that we could not do without them (Norman, 1988), or at least could not do as well in terms of effectiveness, efficiency, and productivity. A good example might be how the recent rapid expansion of distance education is in part related to the affordances that Web-based technologies provide. One set of affordances we find particularly interesting involves multiple modes (e.g., text, audio, graphics) of synchronous and asynchronous communication over great distances, and at any time.

Recently, educational researchers have paid considerable attention to immersive, multi-modal technologies. One of these foci has been on the use of multi-user virtual environments (MUVEs) to support learning across curricula: "educational MUVEs have emerged in recent years as a form of socio-constructivist and situated cognition-based educational software" (Nelson & Ketelhut, 2007, p. 269). There has also been an emerging discussion about MUVEs as tools for second- and foreign-language instruction (e.g., Cooke-Plagwitz, 2008). This phenomenon has likely emerged from a combination of research on Computer Assisted Language Learning (CALL) and Computer Mediated Communication (CMC). Digitally mediated literacy practices (e.g., list-servs, blogs, online chatting, e-mail correspondence, online postings) and social networking practices mediated by such tools as Facebook, MySpace, and Xanga can provide language learners with meaningful opportunities to engage in multiple literacy practices and to construct learner identities through interactive activities in virtual communities (Black, 2005; Lam, 2000; Yi, 2008). Our work is informed by social constructivist principles as applied to foreign-language learning. Our efforts to this point have been to explore how the affordances of Multi-User Virtual Environments (MUVEs) might mediate the learning of English as a Foreign Language (EFL) at a distance.

A primary focus of CALL and CMC research (especially language-learning at a distance) has been on the use of technologies for the development of literacy in a second or foreign language (Lam, 2000; Shei, 2005). Yet, little is known about how emerging technologies, and MUVEs in particular, influence the ways in which language learners improve their oral proficiency. In addition, despite the great potential of this emerging technology to augment language instruction, very little research has been conducted on the use of any form of MUVE in the context of learning English as a Foreign Language (EFL) — where English is not spoken in everyday life, but is often limited to the classroom (e.g., learning English in China).

We chose to explore one of the largest and most well-known MUVEs, Second Life (available at http://secondlife.com). Second Life (often abbreviated as SL) can provide a potentially vast array of rich environments within which users can interact. SL offers friendly, appealing, and contextually relevant spaces, such as offices, shops, athletic events, business meetings, and classrooms for language learners to interact with native speakers of a target language. SL can also offer access to up-to-date online instructional resources in a variety of media formats (e.g., video, photographs, documents, interactive lessons, simulated gaming). In SL, learners can communicate through text messages, audio conversations, and non-verbal gestures (e.g., waving, clapping hands). These types of gestures are performed by their virtual personas or "avatars." An avatar is the 3-D graphic representation through which one interacts in a MUVE. Second Life also offers capabilities to record events taking place within the MUVE. This provides language learners with the opportunity to review and reflect on their virtual experiences. The recording can also provide instructors and researchers with a second chance to analyze their students' performance.

In early 2007, SL introduced Voice over Internet Protocol (VoIP) into its architecture. Prior to the introduction of VoIP, most conversations in SL were conducted through text chat (Au, 2008). VoIP capability has allowed users to communicate verbally in real time, adding a layer of authenticity to the more common text-based interactions. Equally important is the fact that text-based interactions online (i.e., written forms of discourse) have been shown to have significant relationships or overlaps with oral discourse (Belcher & Hirvela, 2008). Moreover, one form of discourse could help language learners improve their skills in the other form. In other words, second-language learners' text-based interactions within MUVEs are likely to help them improve their spoken discourse. Given the features of MUVEs noted above (i.e., realistic, authentic, and relevant settings; VoIP; and the relationship between oral and written discourse), MUVEs present a unique opportunity to practise speaking in a target language with native speakers in authentic contexts. This is particularly valuable for learners who have a very limited opportunity to hear, use, and practise English in real-world and/or offline contexts.

Context

Chinese educators are gradually adopting constructivist perspectives on learning and instruction (Yu, Wang, & Che, 2005), as they are simultaneously seeking emerging technologies that can effectively improve learning and instruction in authentic and meaningful ways. Given this paradigm shift and the potential affordance SL can lend to language learning, we collaborated with a Chinese university to pilot the integration of SL into a speaking course for first-year Chinese EFL students.

YT University (pseudonym) is a major university founded in 1984 in eastern China. The comprehensive university has twenty colleges. YT University's College of Foreign Studies has five departments. The participants in our SL pilot program were first-year EFL students in the English department at YT University. We implemented the pilot in a two-hour weekly speaking class during the spring semester of 2008. The instructor of the course designed activities centred on topics such as globalization with the intent of improving students' English-speaking skills through reading and discussion. We then designed a set of related SL learning activities that would allow the Chinese students to practice what they had learned with native speakers.

This pilot program was implemented online in SL with participants in the language labs both at YT University and at the American university. Thirty-one Chinese EFL students from the YT University English department participated in the SL pilot program. A group of five American graduate students from a large southeastern university also participated as native English-speaking counterparts. Three of the American participants were doctoral students of Instructional Design and Technology, and the other two were graduate students of Applied Linguistics. The American students were all proficient in SL and were asked to read a text on globalization before interacting with their Chinese counterparts. The Chinese participants first completed a survey designed to gauge their technology proficiency. Immediately afterwards, they all participated in a one-hour workshop that familiarized them with basic navigational and communication functions of SL. Each participant was then provided with names and passwords for pre-built avatars in a given SL location. After the workshop, the

Chinese participants interacted with their Chinese peers in SL to both familiarize themselves with the environment and prepare a list of questions about globalization they wanted to ask the American participants.

We next divided the Chinese participants into two groups. Sixteen participants in the first group interacted one-on-one with an American participant to complete the given tasks in SL (four American participants each interacted one-on-one with four different Chinese participants). Fifteen Chinese participants in the second group formed three sub-groups of five to interact with an American participant. All of the participants (both the American and the Chinese) logged into SL using pre-configured avatars. Their avatars had been placed in pre-set locations within SL. Chinese participants and their American counterparts were next given two language tasks. The first task for Chinese participants was to interview an American student about his/her perspectives on globalization in order to write an article for their university newsletter. The second task was for the American participants to interview the Chinese participants about student life in China. The first task was designed to help Chinese students practise oral skills such as questioning and clarification (or information seeking). The second task was designed to help Chinese participants practice listening to and answering questions in the target language.

Before implementation, we collected data on the Chinese students' technology competencies and perceived readiness for the activities. Participants' interactions with one another were observed in real time and were video-recorded using both the recording function embedded in Second Life and Camtasia Studio™. Recording was done from both the Chinese and American sites. Immediately after the completion of the two language-learning tasks, the Chinese participants completed a post-task survey designed to glean participant perspectives on aspects of the learning experiences in SL, including content, activity format, and the SL learning environment. Once the Chinese participants finished the survey, a researcher from the same southeastern American university conducted debriefing interviews with each of the two Chinese groups in English. This data led us to a few important insights into the pilot program, including recommendations for redesign.

Considerations for Integrating Second Life into Language Learning

Task-centered program design

First, we found it was important that the program employ a task-centred approach (Merrill, Barclay, & Schaak, 2008). According to Merrill, Barclay, and Schaak, learning is promoted when learners are engaged in meaningful "task-based" speaking or writing activities (Chapelle, 1998; Ellis, 2003). We would encourage designers to engage learners in meaningful tasks using the target language, thus promoting the use of "authentic language," as opposed to more traditional, de-contextualized grammar instruction or repetitive drills. We also recommend designing a wide array of tasks that can elicit structured interactions (e.g., interviewing native speakers) as well as semi-structured or improvised interactions. Hence, in the future, we would like to add tasks that can lead students to experience impromptu interactions, such as virtual field trips to historically significant places, visits to virtual museums, organizing and participating in virtual conferences, designing and constructing cultural centres, and creating virtual art shows. Activities that involve both EFL students and native-English speaking students in SL could engage all participants in varied meaningful interactions, which would not only equip the EFL students with knowledge about English-speaking cultures, but also provide them with appropriate ways of expressing themselves under certain social circumstances.

In addition to employing both structured and semi-structured task-based activities, we suggest that courses take an "integrative approach" through which students practise both speaking and writing while emphasizing the importance of the relationship between writing and speaking (Belcher & Hirvela, 2008; Weissberg, 2006). Weissberg (2006) argues for the importance of using dialogue (speaking) for second-language (L2) writing, based on his belief in the inextricable link between written and oral modalities. Writing prior to speaking can prepare EFL students with accuracy and fluency of language use in oral interactions while providing them with a purpose for engaging in their oral interactions, and vice versa. It is likely that students may be able to draw upon one relatively stronger skill (e.g., speaking or writing) to support the other weaker skills. Given this potential, it is critical to design a program that can integrate the practice of several language skills in a single experience.

Considering the enabling technology

Another consideration for integrating SL into language learning is the role of an enabling technology (SL in this case) and the significance of interaction that can be mediated by such a technology. According to Vygotsky (1978), social interactions promote learning. Interactions in a 3-D virtual environment, however, require special traits (Jensen, 1999). A language learner in SL interacts through an avatar with other avatars that represent other human beings at the same time, but not within the same physical proximity. In terms of communication and interaction, the interactions undertaken in SL are related to real-life interactions but mediated by SL virtual environments through the Internet and computer applications. Anderson (chapter 2) described the several types of interactions that exist between learners, instructors, learning materials, and environments, while Jensen (1999) has noted several types of interactions that exist specifically in 3-D learning environments. Based in part on these observations and our own, we propose three levels of interaction to consider when engaging students in language learning in SL.

Level 1: Interactions between students and their avatars. Interaction at this level consists of participants controlling and manipulating their avatars in SL. To successfully engage in interactions at this level, participants need to have basic technological competencies, such as controlling a computer keyboard and mouse, and understanding basic navigational functions of the SL virtual environment.

Level 2: Interactions between students and virtual environments. This type of interaction involves participants interacting with various virtual objects in SL. Interactions at this level require students to understand virtual objects in order to communicate within the virtual environment. Searching and saving information, following instructions in a virtual lab, playing a virtual language game, reading virtual books, and using virtual tools for teaching and learning are examples of this type of interaction. In addition to the competencies required in Level 1, to successfully engage in interactions at this level requires participants to understand the navigational systems and the technological functionalities and limits of SL.

Level 3: Interactions between students and other avatars. This type of interaction occurs when students interact and communicate with other avatars in SL through controlling their own avatars. This type of interaction happens in the form of written texts such as chats and instant messages, voice chats, body language (e.g., gestures), and the exchanging of virtual objects including electronic files, html addresses, video clips, and so forth. Chatting with another avatar in a public event, participating in a special-interest group discussion, collaborating with others to build a virtual house, and taking a virtual tour are examples of this type of interaction. This type of interaction is interpersonal and more complex than the interactions at Level 1 and Level 2. To successfully engage in interactions at this level requires students to know how to control their avatars, how to interact with the environment, and how to use the target language in different formats (written, oral, and gestures); to be acquainted with cultural factors such as social rules (e.g., being polite and respectful); and to have personal knowledge of their avatar's identity. We feel that an awareness of these different levels of interactions will help us inform future designs.

Preparing participants

Students need to be well prepared in order to take full advantage of SL. Lim, Nonis, and Hedberg (2006) explored ways in which a MUVE known as "Quest Atlantis" (QA) was used in science lessons to support eight elementary students' learning in Singapore. From pre- and post-tests, interviews, and observations, they observed a low level of engagement in related virtual inquiry activities. This finding was attributed partly to participants' difficulties with the language used in question and answer sessions, and their lack of computer competency.

Data from both the post-program survey and interviews from our project indicated that the participants found their speaking tasks in SL to be relevant and engaging because they had been prepared through preliminary readings and discussions with their Chinese instructors. Although some of the Chinese participants expressed certain anxieties at the beginning of learning tasks in SL, the anxieties appeared to diminish once they were able to work with the familiar content, especially when they found themselves able to interview and discuss globalization with a native-speaking American graduate student. This preparation, as reported in the focus interview after our pilot program implementation, increased participants' confidence in task completion and hence made better use of SL for language practice.

Prensky (2001) uses the term "digital natives" to describe today's students who are "native speakers of technology, fluent in the digital language of computers, video games, and the Internet" (Prensky, 2005, p. 9). Many of the EFL students described in this chapter grew up with computers and the Internet. In addition, they had all taken a required general computer competency course during their first year of study. Although their computer competencies seemed to be high, extra time was needed for them to explore SL. The one-hour training we provided on SL was helpful to get participants started, but more time was needed to prepare them to become comfortable with their avatars and the virtual environment before they engaged in their language activities. For example, when completing learning tasks in SL, the Chinese participants were observed using gestures and text chats in addition to their audio chats. Chinese participants reported in the interviews that their language performance was facilitated by the combination of different modes of communication. These multiple modes of communication should be modelled for students during their MUVE orientation. Two of the Chinese participants in our pilot showed strong dislikes of their avatars and spent a considerable part of their activity time modifying their avatars' appearances. Meadows (2008) found that avatar appearances are important to students partly because "the avatar is a self-portrait" (p. 106). That is, avatar appearances can be important to students either because students feel the need to represent themselves authentically or because students desire the anonymity that comes through avatar representation. We will also provide participants time to design or customize their avatars before they engage in language-learning tasks in future iterations of our project.

Preparing the SL learning environment

First, we suggest that environmental elements both physical and virtual need to be considered. Physical-world constraints, such as Web access and students' physical surroundings, affect students' online experience. For example, we found that computer stations needed to be equipped with large amounts of RAM, high-end graphic cards, and fast processors as the activities in SL using text, audio, and video communications required extensive hardware resources. The computer labs also need to be designed to maximize both collaborative and immersive online experiences. For example, we found that it was helpful for students to be able to record their experiences in SL and reflect on them. This type of video recording usually takes a large amount of hard-drive space. Many language labs in China do not even allow students to save their files on the computer. This leads us to believe that administrative policies need to be more flexible to adjust to pedagogical demands.

Figure 15.1 Students at YT University completing their language tasks in SL

Equally, virtual environments need to be carefully considered. There are many objects in SL that could interfere with and disrupt students when attempting to complete given tasks. Learning experiences should be designed to help avoid unnecessary distractions. For example, it may not be appropriate to select crowded or busy areas for students to meet and perform demanding language tasks. Virtual spaces must also be distinct enough and with enough virtual "distance" between one group of students and the other that voices do not carry from one virtual space to another. While learning spaces were clearly defined and worked adequately during design and testing, when multiple users were logged in and speaking within those spaces, echoes of voices and phantom voices could be heard through the virtual walls of the space. Precisely in the same way that noise management in a physical learning space must be controlled or accounted for during group activities, noise management immediately became a critical issue during our SL experience. In the future, a specific environment designed for the purpose of the study would not only need to include virtual structures and noise barriers where appropriate, but pre-project testing would need to adequately account for the numbers of students speaking simultaneously within a close virtual proximity.

Preparing learning tasks

First, we suggest that clear details on task procedure and desired outcomes need to be provided to the students. Depending on the nature of the tasks that students are required to complete, task procedures and task outcomes may vary, but clear instructions not only provide students with a clear picture of desired outcomes, but they can also provide some motivation for interacting with others in SL.

Our SL activities required the participants to be present in the MUVE simultaneously. Scholars have long recognized time as an important factor in student-language performance (Ellis, 2003; Lee, 2000). To help students be more efficient in their task completion, we strongly recommend that teachers set a time limit for any given tasks in SL, especially when it involves participants across continents.

Facilitators need to be present during their students' initial experiences in SL. Facilitators' presence, as reported in our student interviews, reminded the Chinese students of the importance of the language tasks. Unexpected distractions and interruptions, such as uninvited avatars flying around or falling down right in front of the student avatars, may also have been easier to overcome with the help of an experienced facilitator. We also noted that students found it to be beneficial to use the "Instant Message/Call" communication tool in SL to ask for help during their task completion.

Figure 15.2 SL environment

Ellis (2003) found that post-task reflection is an important part of the task-based learning process. Post-reflection can be done in different formats. Some common post-reflection activities include: (a) individual student reports on task completion (oral or written), (b) a group discussion on task completion, and (c) watching student SL video clips in class and then having students comment on their language performance. The focus of post-task reflection can vary. For example, participants could focus on language-learning targets, such as language accuracy, language fluency, or a particular language use under a certain situated context. We believe that having students watch their own language performance recorded in SL helps with their reflection and can be beneficial to their language-learning experience.

According to Mory (1996), feedback can include a wide array of information, from answer correctness and language uses to motivation messages. The Chinese EFL students in our SL pilot program clearly expressed that they wanted feedback on their language performance from both their Chinese facilitators and their American counterparts. In their own words, "We need to know how well we did it." We strongly recommend providing feedback both during and after task completion. Feedback should include both encouragement and error correction. Thanks to the affordance provided by SL, it is not difficult for teachers to video-record student language performance in SL, even though it might be time-consuming to replay it. This function can help teachers provide accurate feedback on their students' language performance by pinpointing and discussing issues while reviewing events as they (re) occur on the video clips.

Conclusion

In this chapter we discussed our rationale for integrating Second Life into an EFL speaking course in China. Our work was grounded on constructivist learning perspectives and attempted to draw upon the affordance that SL could offer to this context. Finally, we briefly introduced our pilot program and a set of recommendations for future iterations. The rapid evolution of digital media has given us the opportunity to assimilate new technologies while exploring new pedagogies in distance education and language learning. We believe that the work described in this chapter has barely touched upon this potential and we are excited about what the future may bring.

REFERENCES

Au, W.J. (2008). *The Making of Second Life: Notes from the New World*. NY: Harper Collins.

Belcher, D. & Hirvela, A. (Eds.). (2008). *The Oral-Literate Connection: Perspectives on L2 Speaking, Writing, and Other Media Interactions*. Ann Arbor, MI: University of Michigan Press.

Black, R.W. (2005). Access and affiliation: The literacy and composition practices of English language learners in an online fanfiction community. *Journal of Adolescent and Adult Literacy, 49*(2), 118–128.

Chapelle, C. (1998). Research on the use of technology in TESOL: Analysis of interaction sequence in Computer-Assisted Language Learning. *TESOL Quarterly, 32*(4), 753–757.

Cooke-Plagwitz, J. (2008). New directions in CALL: An objective introduction to Second Life. *CALICO Journal, 25*(3), 547–557.

Ellis, R. (2003). *Task-Based Language Learning and Teaching.* Cambridge: Oxford University Press.

Jensen, J.F. (1999, 24–30 October). *3D-inhabited virtual worlds: Interactivities and interaction between avatars, autonomous agents, and users.* Paper presented at the WebNet 99 World Conference on the WWW and Internet, Honolulu, Hawaii.

Lam, E.W.S. (2000). L2 literacy and the design of the self: A case study of a teenager writing on the Internet. *TESOL Quarterly, 34*(3), 457–481.

Lee, J. (2000). *Tasks and Communicating in Language Classrooms.* Boston: McGraw-Hill.

Lim, C., Nonis, D., & Hedberg, J. (2006). Gaming in a 3D multi-user virtual environment: Engaging students in science lessons. *British Journal of Educational Technology, 37*(2), 211–231.

Meadows, M.S. (2008). *I, Avatar: The Culture and Consequences of Having a Second Life.* Berkeley, CA: New Riders.

Merrill, M.D., Barclay, M., & Schaak, A.V. (2008). Prescriptive principles for instructional design. In M. Spector, D. Merrill, J.V. Merrienboer, & M. Driscoll (Eds.), *Handbook of Research on Educational Communications and Technology* (3rd ed.) (pp. 523–556). New York, NY: Lawrence Erlbaum Associates.

Mory, E.H. (1996). Feedback research. In D.H. Jonassen (Ed.), *Handbook of Research for Educational Communications and Technologies.* New York, NY: Simon & Schuster MacMillan.

Nelson, B.C., & Ketelhut, D.J. (2007). Scientific inquiry in educational multi-user virtual environment. *Educational Psychology Review, 91*(3), 265–283.

Norman, D.A. (1988). *The Psychology of Everyday Things.* New York, NY: Basic Books.

Prensky, M. (2001). Digital natives, digital immigrants. *On the Horizon, 9*(5), 1–2.

Prensky, M. (2005). Listen to the natives. *Educational Leadership, 63*(4), 8–13.

Rosenburg, M. (2001). *E-learning: Strategies for Delivering Knowledge in the Digital Age.* New York, NY: McGrawHill.

Shei, C. (2005). Integrating content learning and ESL writing in a translation commentary writing aid. *Computer Assisted Language Learning, 18*(1–2), 33–48.

Vygotsky, L. (1978). Interaction between learning and development. In M. Gauvain & M. Cole (Eds.), *Readings on the Development of Children* (pp. 34–40). New York, NY: Scientific American Books.

Weissberg, R. (2006). *Connecting Speaking and Writing in Second Language Writing Instruction.* Ann Arbor, MI: University of Michigan Press.

Welsh, E.T., Wanberg, C.R., Brown, K.G., & Simmering, M.J. (2003). E-learning: Emerging uses, empirical results, and future directions. *International Journal of Training and Development, 7*(4), 245–258.

Yi, Y. (2008). Relay writing in an adolescent online community. *Journal of Adolescent & Adult Literacy, 51*(8), 670–680.

Yu, S., Wang, M., & Che, H. (2005). An exposition of the crucial issues in China's educational informatization. *Educational Technology Research and Development, 53*(4), 88–101.

16 ANIMATED PEDAGOGICAL AGENTS AND IMMERSIVE WORLDS:

Two Worlds Colliding

> *Bob Heller & Mike Procter*

Abstract

Animated pedagogical agents (APAs) have a number of potential benefits that are especially relevant to distance education, including improved communication and increased student motivation/engagement. Yet, the literature on these benefits is equivocal due in part to the limited range of applications. A review of APA taxonomies reflects the influence of intelligent tutor systems (ITSs) but also reveals a number of important distinctions. One type of APA is identified, actor agents, that may be particularly useful for distance educators, especially if these agents are given conversational abilities and a "stage" on which to perform, such as Second Life (SL). Several projects involving actor agents and SL are described with a conclusion that actor agents in immersive worlds are an opportunity to be grasped.

Introduction

One of the goals of this chapter is to review the literature on animated pedagogical agents (APAs) and how these findings can be effectively applied to distance education. Clearly, understanding the benefits of emerging technologies in general, and APAs in particular, especially how they apply within the different types of APA applications, is an important prerequisite in the design of effective APAs for distance education (see chapter 3). A second goal of this chapter is to explore the role of immersive environments in providing a digital space for APAs to operate. An immersive environment can be defined as a computer-created scene or "world" within which a user can immerse him or herself and interact with in-world objects and other users as an avatar, or computer-generated character. Although there are many different applications of immersive environments in education, our focus is on how the performance of conversational actor agents, a type of APA, could be significantly enhanced by embedding them in virtual environments for interaction with other users and/or agents.

In the next three sections, the hypothesized benefits associated with APAs are reviewed (especially as they apply to distance education), various APA taxonomies are presented, and pilot work on historical figure APAs and distance education is described. The final two sections provide a brief review of immersive worlds in education with a focus on Second Life (SL), and how worlds such as SL can provide a programmable stage for actor agents working with distance learners.

Animated Pedagogical Agents

Animated pedagogical agents can be defined as animated computer-generated characters that respond to user input, adapt to user behaviour, and facilitate learning in a computer-based learning environment (Johnson, Shaw, & Ganeshan, 1998). Given the growing role of course management systems and e-learning in distance education (Holmes & Gardner, 2006; Stewart, Gismondi, Heller, Kennepohl, McGreal, & Nelson, 2007), it is important to examine the claims surrounding the benefits of APAs and the possible role that APAs could play in distance education.

Johnson, Rickel, and Lester (2000) argue that APAs were created when animated interface agents were combined with intelligent tutor systems (ITSs), the de facto standard of computer-based learning environments. According to Johnson et al., APAs provided two key advantages over previous work. First, APAs provided an opportunity to increase the "bandwidth of communication" between students and computers. Through the use of gaze, gesture, and other paralinguistic cues, animated agents can transmit information with greater fidelity and clarity, which in turn can improve learning transfer and outcomes. Second, Johnson et al. (2000) and others (Lester et al., 1997) argue that

APAs increase student engagement and motivation with regards to the learning task. Lester et al. coined the phrase "persona effect" to describe the increase in student motivation and engagement associated with the use of APAs in computer-based learning environments. According to Lester et al., the mere presence of an animated agent had a strong positive effect on learner perceptions of the learning experience. These two benefits are especially important to note since effective communication and student motivation/engagement are traditionally problematic areas in distance education.

Gulz (2004) also noted the motivational and communicative benefits proposed by Lester et al. (1997) and identified four additional benefits associated with the use of APAs in computer-based learning environments. Gulz (2004) suggested that APAs can increase the sense of ease and comfort and fulfill a need for a personal connection in the learning task. Collectively, these two benefits contribute to the sense of social presence experienced by the user. Social presence is an important construct in distance education and can be defined as the ability of learners to be socially and emotionally connected to a community of inquiry (Garrison, Anderson, & Archer, 2000). Social presence supports learning by appealing to the essential social nature of human beings. Learning is sustained with increased persistence when interactions are "socialized" to be engaging and appealing. Social presence is closely related to Reeves' and Nass' (1996) media equation, which states that "individuals' interactions with computers, television, and new media are *fundamentally social and natural,* just like interactions in real life" (p. 5). Social presence is an important outcome measure in any APA evaluation and should be reflected, if it exists, in the conversational record left behind from the learner-APA interaction.

Interestingly, the final two benefits listed by Gulz (2004) also can be contrasted with the remaining constructs from Garrison et al.'s (2000) Community of Inquiry model. Specifically, Gulz refers to the potential of APAs to stimulate essential learning activities, such as exploration, attending, and reflection — activities that seem related to the construct of teaching presence. Teaching presence can be defined as the systematic management of cognitive and social processes necessary to achieve desired learning outcomes. The last benefit described by Gulz suggests improved cognitive outcomes may be realized by using APAs in the areas of memory, problem solving, and comprehension. This cognitive outcome is consistent with Garrison et al.'s (2000) cognitive presence construct, which can be defined as the extent to which learning (i.e., meaning construction) occurs following sustained communication. Overall, it seems the benefits listed by Gulz (2004) fit well with the Community of Inquiry model.

In sum, the benefits of APAs as stated are potentially enormous, especially as they apply to the challenges in distance education. However, their actual impact has been limited, and evidence for their putative effects has been equivocal at best (see Clark & Choi, 2005; Dehn & van Mulken, 2000; Gulz, 2004; Gulz & Haake, 2006). Dehn and van Mulken (2000) concluded that evidence for a persona effect is weak and confined primarily to affective self-report measures. Clark and Choi (2005) noted that much of the extant literature is weak in both internal and external validity. Gulz and Haake (2006) note a dearth of research on the visual properties of agent representation, in spite of the broader literature on visual effects on cognition. Finally, among the reasons given for these null findings, is the suggestion that most of the existing APA research has been restricted to a limited range of applications that focus primarily on instructional roles (Clarebout, Elan, & Johnson, 2002; Gulz, 2004; Payr, 2003). In the next section, taxonomies for categorizing agents are presented with a focus on applications most suited for distance education.

Types of APAs

One of the earliest attempts to classify APAs was provided by Baylor (1999), who first argued that intelligent agents are best conceptualized as "cognitive tools" that can be used by students to support, guide, and extend their thinking processes. This approach stands in contrast to the "intelligent tutor" approach, which is focused on modelling effective tutor behaviour and uses technology to constrain student learning. As cognitive tools, Baylor suggests three types of educational applications for intelligent agents: information managers, pedagogical experts, and programmable learners. Information managers support learning by reducing the cognitive load associated with the processing of excess information. Pedagogical experts monitor and evaluate the timing and implementation of pedagogical strategies as they work with students learning to master domain-specific knowledge. Programmable learners are agents who enable "students creating agents" to be part of the pedagogical strategy.

Chou, Chan, and Lin (2003) offer a similar set of distinctions for educational agents: personal assistants, pedagogical agents, and learning companions. Personal assistants are much like information managers, and pedagogical agents seem identical to pedagogical experts. Learning companions are an expansion of the programmable agent and can take many forms (competitor, collaborator, tutee, peer tutor, trouble maker, critic, or clone), but they are essentially there in a non-authoritative role to create a social dynamic that supports learning. Furthermore, like programmable learners, the roots of learning companions lie in the "learning by teaching" approach (Ur & Van Lehn, 1995).

Clarebout et al. (2002) developed a typology based on the instructional role or pedagogical strategy of an agent (i.e., supplanting, scaffolding, demonstrating, modelling, coaching, testing) and its modalities of support (executing, showing, explaining, and questioning). In an analysis of over twenty agents, they report that most APAs act as coaches and provide content and problem-solving support. They also note that there were few APAs that focused on providing metacognitive support, in part because of the association of APAs with ITSs. Most ITSs are focused primarily on the acquisition of domain-specific knowledge within a well-defined problem space where a single solution exists. The intelligence resides in modelling a good tutor and knowing where the student is located in the problem space. Clarebout et al. (2002) suggest that analyzing student behaviour, rather than tutor behaviour, may reveal a stronger need for metacognitive support to help students monitor and manage their own learning process. Clarebout et al.'s (2002) suggestion is similar to Baylor's (1999) assertion that intelligent agents are best conceptualized as cognitive tools to support metacognitive processes. Kerly, Hall, and Bull (2006) also believe that support for metacognitive processes is important in open learner modelling and specifically, conversational agents employing natural language can help users negotiate a model of their own understanding.

Payr (2003) categorizes educational agents into tutors, coaches, agents to support collaboration, learning companions, and agents as actors. Payr's definition of learning companions is more restrictive than Chou et al.'s (2003) but her taxonomy does parse out an actor agent type of APA that is closer to Baylor's notion of a programmable agent. However, the role of an actor agent seems much broader because in addition to being taught by students, actor agents can participate in a wide range of pedagogically informed simulations and replications. According to Payr (2003), this type of agent system is the most interesting of all agent types and could be used in training simulations for micro-level social interactions, which are crucial components in all professions with human-to-human services. Payr (2003) laments the fact that much of the research on educational agents uses "new technology for old learning" and argues that new forms of learning are possible when users are allowed to freely interact with synthetic characters.

One final dimension should be noted regarding the classification of agents. Veletsianos, Scharber, and Doering (2008) make a distinction between pedagogical agents in terms of conversational ability. They argue that interactive conversational ability is a critical feature that directly impacts the student experience, the way in which students interact with agents, and their perceptions and knowledge gained as a result of those interactions. Certainly the work of Cassell and colleagues on embodied conversational agents is a testimony to the singular importance of conversation and its role in communication (Cassell, Bickmore, Campbell, Vilhjalmsson, & Yan, 2000). Kerly et al. (2006) also support the role of conversational agents for using natural language to negotiate a learner model. In this regard, it is interesting to note that work in intelligent tutor systems is now recognizing the importance of narrative as a means to enhance motivational effects and improve learning outcomes (e.g., McQuiggan, Rowe, & Lester, 2008). Finally, conversational agents allow users to project their sense of social presence into the conversational flow.

In summary, a number of taxonomies have been proposed for classifying APAs with different roles that can be examined in relation to the hypothesized benefits associated with APAs. However, it is important to note that the vast majority of APAs fall under the coach/tutor role, reflecting the essential component of ITS design (Clarebout et al., 2002). Given the importance of engagement and motivation as APA benefits, we believe that agents as actors afford the greatest opportunity to maximize engagement and motivation. For example, Veletsianos and Miller (2008) state that virtual historians could be designed that motivate students to examine historical events and concepts. They suggest that digital representations of historical figures (e.g., Dwight Eisenhower, Winston Churchill, and Nelson Mandela) could be used to engage students in meaningful interactions about past events and personalities. In the next section we elaborate on the rationale for historical figures as actor agents, including a brief description of our own research. The following section argues that the emergence of immersive worlds may be the stage that enables actor agents to truly perform.

Historical Figure Applications

As noted earlier, the APA's ability to engage and motivate the student is a critical prerequisite for the persona effect to occur (Gulz & Haake, 2006). Like Veletsianos and Miller (2008), we believe that actor agents based on well-known historical figures may generate significant intrinsic interest in the users and, in turn, maximize the APA's potential to engage and motivate. Moreover, users with an intrinsic interest in the APA may adopt a lower psychological threshold for agent believability and realism. Johnson et al. (2000) noted that, like agents designed for entertainment, APAs must be lifelike and believable in order to maximize engagement. These are critical aspects of APA design that tend to require costly technology.

Unlike the role a student adopts when interacting with a tutor/coach APA with structured problem-solving and procedural-like solutions, the role of a learner interacting with a historical figure actor agent is more interactive, like that of a journalist or interviewer. The learner's primary task is to formulate questions and comments around the life and times of the historical figure and contribute appropriately to the ongoing conversation. The nature of the application encourages students in this role to explore topic areas and reflect on the responses provided in order to construct meaning from their interaction.

In response, a historical figure actor agent should be prepared, at the very least, to answer domain-relevant questions as well as questions that are more autobiographical in nature. Moreover, historical figure actor agents should be prepared to provide responses in the form of a narrative, which is expressed using the turn-taking rules and expressions associated with effective conversation. Many of our day-to-day conversations involve the communication of the episodic events that occur in our lives. Some of these events are celebrated narratives from our past, whereas others may revolve around the more mundane events of daily living. Surprisingly, the use of narratives by APAs is very infrequent despite the arguments supporting narratives as effective pedagogical tools for social exchange and learning (Heo, 2004; Shank, 1995). Thus, the conversational ability and narrative capacity of a historical figure actor agent will be a critical feature that determines success.

Not only do historical figure applications increase the range of social roles and provide opportunity for the use of narratives, they also provide a tighter synthesis between content and persona. As Johnson et al. (2000) noted, many APAs are the combination of an animated interface agent attached to the front end of an intelligent tutor system, and evaluations of the agent can also include evaluations of the system (or be driven by system performance). In a historical figure application, the content or system is more tightly woven into the agent.

In sum, we argue that historical figure actor agents are a strong test of the engagement function of an APA, which should maximize the hypothesized persona effect and other benefits associated with APAs. In addition, the historical figure application stresses the importance of conversation and narrative as the basis of information exchange and extends the set of social roles for students in learning and content interaction.

To investigate this type of actor agent application, we developed

Freudbot: a historical figure agent based upon Sigmund Freud, arguably the most well known figure in psychology. Conversational ability was modelled using Artificial Intelligence Markup Language (AIML), an XML-based open-source programming language, developed by Richard Wallace. AIML is the language used to support ALICE (http://www.alicebot.org), an award-winning chatbot and progenitor of thousands of other chatbots as hosted on Pandorabots (http://www.pandorabots.com). At its core, AIML relies on pattern matching and consists of "category" elements that in turn contain a "pattern" and "template" elements. If the input matches the pattern, the template defines the action to be taken. Simple categories can be combined using a built-in recursion function. Logic flow can be achieved using basic conditional operations, which are also part of the AIML language. Adding content to AIML agents is an iterative and incremental manual process where user input is targeted for failed matches and new content is added.

In addition to programming content, AIML was used to manage the dialogue output in the form of narratives similar in function to the story grammar approach of Thorndyke (1977). Freudbot's primary content was represented as ninety-one autobiographical and conceptual narratives that in turn were composed of three to seven narrative sections. When users typed specific keywords and phrases, Freudbot would provide a section with implicatives that would invite the user to request more information using conversational directives (e.g., go on, tell me more, is that all, why is that, etc.), a feature consistent with the conversational rules related to turn-taking. The learner can effectively control the way in which a story can be told by switching to other stories or entering into specific parts of a story. Freudbot also has the capacity to return to a story after branching to a new location and also retains the parts of a story that have been told to prevent repetition. We also developed agent strategies loosely consistent with speech act theory. In cases in which no input was recognized, the agent would default to one of several conditional strategies: ask for clarification, suggest a new topic for discussion, indicate that he had no response, or ask the user for a suggested topic.

A proof of concept study was carried out in which fifty-three students in psychology chatted with Freudbot for ten minutes and then completed a questionnaire that collected information on the learning experience and other relevant demographic variables (see Heller, Procter, Mah, Jewell, & Cheung, 2005). Approximately 68 percent of the students indicated they would chat again with Freudbot, and of these students, a composite measure of their conversational experience was significantly higher than the midpoint of the scale from which it was drawn (i.e., significantly higher than three on a five-point scale). Moreover, when students were asked to rate various conversational agent applications for their utility, historical figures applications were the highest-rated application and rated significantly higher than course administrative agents, course content agents, and chatroom agents.

More recently, the findings from above were replicated in a second study involving Freudbot that also examined the persona effect in response to different image conditions (Heller & Procter, 2009). Surprisingly, we found that significantly higher ratings for the learning experience were reported in the no-image condition than in a static-image condition and an animated-image condition. Although somewhat counterintuitive, the findings may reflect the importance of getting the animation "right." Dirkin, Mishra, and Altermatt (2005) reported a similar finding and called it an "all or none effect." That is, learning outcomes were better in the absence of an agent and in the presence of a social agent in comparison to a nonsocial agent. We are currently running a study to replicate this finding with greater precision on the underlying causal variables. In any event, the findings draw attention to the role of the computer-based environment in which the actor agents are embedded.

Second Life: A Stage for Actor Agents

APAs typically do not operate in isolation. According to Lester et al. (1997), APAs require a computer-based interactive learning environment within which to operate. The majority of these learning environments emerged out of ITSs and as such, were content-specific and tailored to the task at hand. This interface between user and agent is becoming increasingly important. As indicated earlier, agent characteristics are closely related to the emergence of a persona effect. It seems equally important that immediate context of the interaction would influence the persona effect and contribute overall to a sense of social presence.

In this regard, it is important to note the growth of immersive environments in general, and Second Life (SL) in particular. Immersive environments can be defined as a computer-created scene or "world" within which a user can immerse him or herself as an avatar and interact with other users/avatars and in-world objects. Our focus on SL reflects the large user group of educators (the SL educators' listserve has an estimated five thousand users) and the uptake of SL by numerous institutions of higher education. For example, according to a blog on SL, seventeen of the top twenty institutions in the U.S. have established a presence on SL (http://blog.secondlife.com/2008/07/24/my-first-two-months-at-linden-lab/) and a Spring 2008 snapshot of higher education in the UK estimates that three quarters of UK institutions are actively developing in SL (Kirriemuir, 2008). It is especially important to note that SL supports user-created content, and a number of educators have created a wide range of simulations and replications (see http://sleducation.wikispaces.com/educationaluses for a summary of educational applications of SL). An example of the use of SL in distance education is presented in chapter 15.

Surprisingly, the use of actor agents or bots (computer-controlled virtual agents) in SL educational applications is almost non-existent (Ullrich, Bruegmann, Prendinger, & Ishizuka, 2008). This may be due in part to the negative reputation of bots with respect to commercially driven applications in SL, where bots are often employed to inflate a region's visitor count. Since the choice to visit a region is often related to the presence of other visitors, some regions will often use bots to portray an illusion of activity. Bots are also associated with illegally copying users' inventories and supplanting valid users from "camping," a common method of obtaining in-world currency. It has been estimated that up to 20 percent of the users in SL at any one time are actually bots (http://www.massively.com/2008/04/28/peering-inside-how-many-bots/). The absence of actor agents in SL is also unusual given the role and importance of Non-Playing Characters in the gaming world. Even the "holodeck" from the popular Star Trek Enterprise series was known for replicating characters as well as environments.

Finally, the lack of actor agents in SL is surprising given the availability of programming interfaces that enable avatars to act autonomously under computer control. Second Life provides an official scripting mechanism, Linden Scripting Language (LSL), with over three hundred library functions that allow control over and communication between objects and avatars. Linden Scripting Language has several built-in safeguards, in the form of restricted functionality and delays, intended to prevent inappropriate or illegal behaviour by scripted in-world objects. Friedman, Steed, and Slater (2007) have successfully used LSL to create bots that wander around in SL recording data on the spatial social behaviour that arises from chance encounters with other avatars. There also exists an unofficial open-source library, libopenmv (formerly libsecondlife), which provides direct access to SL functions through applications written in C#. Libopenmv gets around restrictions built into LSL, allowing for more sophisticated bot applications. According to Ullrich et al. (2008), the advantages of libopenmv over LSL include more control over avatar behaviour, immediate responses (no delays), and no memory constraints on script size.

Perhaps the most compelling argument for the absence of actor agents in SL lies in the challenges associated with programming to interact in such an unrestricted and often unpredictable environment. Something as simple as determining whether another avatar is speaking to you or someone else requires significant processing (determining and analyzing the proximity of surrounding avatars based on 3-D coordinates and the direction they are facing, and possibly searching the text of their messages for cues that they are addressing you). Currently these functions must be coded manually using relatively low-level functions, whether one is using LSL or libopenmv. Significant increases in the development of actor agents in SL may come when higher-level software routines become available to support functions associated with avatar behaviour, such as navigating around obstacles, or recognizing what other avatars are doing around you.

In sum, although challenging, SL has the potential to be an effective stage for the operation of actor agents. In the final section, we describe some of the preliminary work that is being done to create actor agents in SL.

Actor Agents in Second Life

A prototype simulation involving Freudbot was developed using LSL and implemented on the island owned by Athabasca University. Currently, Freudbot dwells in an office sitting beside a couch. When a visitor enters the room, he or she is given a notecard that describes Freudbot and his purpose. When the visitor approaches Freudbot, he stands up and asks the visitor to sit on the couch if he or she would like to chat. If the user is agreeable and sits on the couch, Freudbot will ask the user what he or she would like to talk about. At this stage of development, Freudbot simply stands and sits down in response to user behaviour, but plans are in place to endow Freudbot with converstaional behaviours that should contribute to a sense of social presence. On Athbasca Univeristy's island in Second Life, Freudbot has approximately one or two visitors per week. Freudbot is also available on The Theorist Project (http://slurl.com/secondlife/MOntclair%20State%20CEHSADP/78/203/23), a build by Montclair State University that is devoted exclusively to the major counseling theorists in psychology. In this context, it is interesting to note that the visits increase significantly to one or two per day. Plans are in place to create similar spaces for Piagetbot, as well as other historical figure actor agents in various stages of development.

Scott Overmeyer from Baker College is developing an impressive simulation of a small town made up of several businesses (grocery store, doctor's office, post office, hardware store, small manufacturing facility). Students can interact with a number of actor agents, which play such parts as shopkeepers, bankers, and real estate agents. For example, for an exercise that entails producing an RFP for an inventory control system, a typical real-world task, a student is able to ask the store manager actor agent questions such as "How do you process a sale?" Like Freudbot, the conversational abilities are based on AIML, using LSL and Libopenmv to control the agents.

Finally, there are a number of SL medical simulations with actor agents behaving as patients (see Imperial College, London: http://www.elearningimperial.com/index.php?option=com_content&task=view&id=37&Itemid=58, and the University of Auckland: http://slenz.wordpress.com/2008/10/26/the-slenz-update-no-19-october-26-2008/). Interestingly, Payr (2003) made this suggestion over five years ago as an example of how actor agents could be used in innovative ways as APAs. However, the focus of these simulations is somewhat structured and the interactivity of the actor agent patients is very limited, with little or no conversational ability. We are currently working with collaborators Doug Danforth (Ohio State University) and Mary Johnson (Florida State University) to develop a virtual patient with conversational abilities based on AIML. Ideally, the patient would be used by first- and second-year medical students to help them practise their clinical interviewing skills.

Actor Agents in SL: An Opportunity to Be Grasped?

In 2002, Clarebout et al. argued that APAs were an opportunity to be grasped as a means of enhancing the use of support tools in computer-based open learning environments. We believe that this opportunity is even greater for actor agents in an immersive world. As Gulz and Haake noted in their 2006 review, the question regarding APAs has moved away from "Do they work?" to "When do they work and in what context?" The work described above is an attempt to answer these questions.

REFERENCES

Baylor, A.L. (1999). Intelligent agents as cognitive tools for education. *Educational Technology, 39*(2), 36–40.

Cassell, J., Bickmore, T., Campbell, L., Vilhjalmsson, H., & Yan, H. (2000). Human conversation as a system framework: Designing embodied conversational agents. In J. Cassell, J. Sullivan, S. Prevost, & E. Churchhill (Eds.), *Embodied Conversational Agents* (pp. 29–63). Cambridge: MIT Press.

Chou, C., Chan, T., & Lin, C. (2003). Redefining the learning companion: The past, present, and future of educational agents. *Computers and Education, 40*, 255–269.

Clarebout, C., Elan, J., & Johnson, W.L. (2002). Animated pedagogical agents: An opportunity to be grasped? *Journal of Educational Multimedia and Hypermedia, 11*(3), 267–286.

Clark, R.E., & Choi, S. (2005). Five design principles for experiments on the effects of animated pedagogical agents. *Journal of Educational Computing Research, 32*(3), 209–225.

Dehn, D.M., & van Mulken, S. (2000). The impact of animated interface agents: A review of the empirical literature. *International Journal of Human-Computer Studies, 52*, 1–22.

Dirkin, K.H., Mishra, P., & Altermatt, E. (2005). All or nothing: Levels of sociability of a pedagogical software agent and its impact on student perceptions and learning. *Journal of Educational Multimedia and Hypermedia, 14*(2), 113–128.

Friedman, D., Steed, A., & Slater, M. (2007). Spatial social behavior in Second Life. *Lecture Notes in Computer Science, 4722*, 252–263.

Garrison, D.R., Anderson, T., & Archer, W. (2000). Critical inquiry in a text-based environment: Computer conferencing in higher education. *The Internet and Higher Education, 2*(2–3), 87–105.

Gulz, A., (2004). Benefits of virtual characters in computer-based learning environments: Claims and evidence. *International Journal of Artificial Intelligence in Education, 14*, 313–334.

Gulz, A., & Haake, M. (2006). Design of Animated Pedagogical Agents: A look at their look. *International Journal of Human-Computer Studies, 64*(4), 322–339.

Heller, B., & Procter, M. (2009). Animated pedagogical agents: The effect of visual information on a historical figure application. *International Journal of Web-based Learning and Teaching Technologies, 4*(1), 54–65.

Heller, R.B., Proctor, M., Mah, D., Jewell, L., & Cheung, B. (2005). Freudbot: An investigation of chatbot technology in distance education. In P. Kommers & G. Richards (Eds.), *Proceedings of the World Conference on Educational Multimedia, Hypermedia and Telecommunications, 2005* (pp. 3913–3918). Chesapeake, VA: AACE.

Heo, H. (2004). Story-telling and retelling as narrative inquiry in cyber learning environments. In R. Atkinson, C. McBeath, D. Jonas-Dwyer, & R. Phillips (Eds.), *Beyond the Comfort Zone: Proceedings of the 21st ASCILITE Conference*, Perth, AUS.

Holmes, B., & Gardner, J. (2006). *E-Learning: Concepts and Practice*. London: Sage Publications

Johnson, L., Shaw, E., & Ganeshan, R. (1998). Pedagogical agents on the Web. Retrieved from http://www.isi.edu/isd/ADE/papers/its98/ITS98-WW.htm

Johnson, W.L., Rickel, J.W., & Lester, J.C. (2000). Animated pedagogical agents: Face-to-face interaction in interactive learning environments. *International Journal of Artificial Intelligence in Education, 11*, 47–78.

Kerly, A., Hall, P., & Bull, S. (2006) Bringing chatbots into education: Towards natural language negotiation of open learner models. *Proceedings of AI-2006, 26th SGAI International Conference on Innovative Techniques and Applications of AI*, Cambridge, UK.

Kirriemuir, J. (2008). A Spring 2008 "snapshot" of UK higher and further education developments in Second Life. Eduserv Foundation.

Lester, J.C., Converse, S.A., Kahler, S.E., Barlow, S.T., Stone, B.A., & Bhogal, R.S. (1997). The persona effect: Affective impact of animated pedagogical agents. In S. Pemberton (Ed.), *Human Factors in Computing Systems: CHI'97Conference Proceedings*. New York, NY: ACM Press.

McQuiggan, S.W., Rowe, J., & Lester, J.C. (2008). The effects of empathetic virtual characters on presence in narrative-centered Learning Environments. In *Proceedings of the 2008 SIGCHI Conference on Human Factors in Computing Systems* (pp. 1511–1520). Florence, Italy: ACM.

Payr, S. (2003). The virtual university's faculty: An overview of educational agents. *Applied Artificial Intelligence, 17*, 1–19.

Reeves, B., & Nass, C. (1996). *The Media Equation: How People Treat computers, Television, and New Media like Real People and Places*. Stanford, CA: CLSI Publications.

Shank, R.C. (1995). *Tell Me a Story: Narrative and Intelligence*. Evanston, IL: Northwestern University Press.

Stewart, B., Briton, D., Gismondi, M., Heller, B., Kennepohl, D., McGreal, R., & Nelson, C. (2007). Choosing Moodle: An evaluation of Learning Management Systems at Athabasca University. *International Journal of Distance Education Technologies, 5*(3).

Thorndyke, P. (1977). Cognitive structures in comprehension and memory of narrative discourse. *Cognitive Psychology, 9*, 77–110.

Ullrich, S., Bruegmann, K., Prendinger, H., & Ishizuka, M. (2008). *Lecture Notes in Computer Science, 4722*, 281–288.

Ur, S., & Van Lehn, K. (1995). STEPS: A simulated, tutorable physics student. *Journal of Artificial Intelligence in Education 6*(4), 405–435.

Veletsianos, G., & Miller, C. (2008). Conversing with Pedagogical Agents: A phenomenological exploration of interacting with digital entities. *British Journal of Educational Technology, 39*(6), 969–986.

Veletsianos, G., Scharber, C., & Doering, A. (2008). When sex, drugs, and violence enter the classroom: Conversations between adolescents and a female pedagogical agent. *Interacting with Computers, 20*(3), 292–301.

CONCLUSION

> *George Veletsianos*

Notwithstanding important global events that occurred between July 2008 and October 2009, the period in which this book was developed (such as the worldwide economic recession and the election of Mr. Barack Obama to the U.S. presidency), technological advances during this time have been rapid. To cite a few, Twitter became part of the popular discourse, the Web has seen increased activity and interest in real-time access to published information, and augmented-reality and location-aware applications are gaining traction. In addition, this period has seen advances in the educational front. For instance, Open Access Week was first celebrated in 2009 with calls for immediate and free access to scholarly knowledge, while two free online universities were launched (Peer-to-Peer University and the University of the People). It seems that both the Web and the way we think about education are changing.

Regardless of the emergent state in which education and the Web are situated, this book provides evidence that we are moving towards a consensus with regards to how effective and engaging learning experiences should be designed. Whether as a result of technological advancements, a changing mindset, or a combination of the two, distance learning educators, researchers, and practitioners are collectively focusing their attention around a recurring theme. Specifically, they are seeking approaches grounded upon social, authentic, and community-based learning experiences, where presence, communication, interaction, and collaboration are valued. In this context, emerging technologies are used to enhance education, and good practice and pedagogy are used to appropriate the emerging technologies available.

Reflecting on the finished chapters, the original submissions, and my discussions with chapter authors, I see three themes that can bring closure to this volume: (a) the focus of the book, (b) the excitement and motivation displayed by this volume's practitioners and researchers, and (c) the prospects for future research. I discuss these themes next.

First, while our focus lies on the use of emerging technologies in distance education, it is clear from reading the chapters and observing the summary of the chapters generated via wordle.com (Figure C.1) that the focus isn't necessarily the technology. The authors in this volume focus on enhancing educational research and practice based on the notion that powerful learning experiences are social, immersive, engaging, and participatory. In turn, these types of learning experiences lend themselves well to being enhanced through the emerging technologies that we have available at our disposal.

Figure c.1 A summary of the sixteen chapters included in this volume (generated on www.wordle.com)

Second, the authors contributing to this volume have displayed tremendous excitement for their work, eagerness to receive feedback, and motivation to transform distance education. These authors are not just scholars but also activists in furthering meaningful, just, and powerful educational opportunities. To me personally, this is very important. The work of an academic should not be limited to teaching classes and writing research reports to be read and analyzed by like-minded individuals. In short, academics should also see themselves as changemakers, and academics in schools of education in particular should focus their work towards developing equitable societies that are free of injustices, where opportunities for deeply personal and powerful learning experiences are open to everyone. Evidence of these authors' commitment to the noble causes of education was the fact that submissions to this book came as a direct result of it being open access. In particular, more than three quarters of the original sixty-five submissions noted that the reason for submitting to this project was because the book was going to be offered free of charge for anyone to use and download.

Finally, while each chapter suggests future lines of inquiry at the micro level, the work presented in this volume collectively highlights broader areas of interest that are worthy of research attention. At the macro level, it is clear that we need longitudinal research that is multidisciplinary in nature. At the meso level, important areas of inquiry and research include:

> further inquiry into the symbiotic and reinforcing relationship between emerging technologies, pedagogies, and the rise of the participatory Web;
> new pedagogies and approaches that embrace emerging technologies as natural artifacts in contemporary educational systems, as opposed to add-ons to an existing pedagogy, approach, or activity;
> renewed emphasis on the role and nature of education and universities, along with an examination of the roles of educators and informal learning experiences;
> further research into understanding how social, immersive, engaging, and participatory learning experiences can be initiated in distance education contexts;
> development of research frameworks for investigating social, immersive, engaging, and participatory learning; and
> revamped efforts to understand how learning communities can be fostered (both in the context of formal education, as well as in the context of lifelong informal learning).

In closing, I hope you enjoyed reading this book and that you found it worthwhile for your research and practice. If you did, let me know at veletsianos@gmail.com and feel free to share the book openly and freely.

George Veletsianos

May 2010

CONTRIBUTORS

Terry Anderson is Professor and Canada Research Chair in Distance Education at Athabasca University—Canada's Open University. He has published widely in the area of distance education and educational technology, and has co-authored or edited six books and numerous papers. Terry is active in provincial, national, and international distance education associations, and is a regular presenter at professional conferences. He teaches educational technology courses in Athabasca's Masters and Doctorate of Distance Education programs.

Trish Andrews is a lecturer in Higher Education in the Teaching and Educational Development Institute (TEDI) at the University of Queensland, Australia. Trish has extensive experience in supporting innovative curriculum development, including integrating technologies into higher education programs. She has been involved in numerous technology innovation activities and has had several educational development and research grants, including a number in the area of videoconferencing and rich media. Trish has a particular interest in mobile technologies and is currently involved in three grants utilizing mobile technologies to address different teaching and learning issues. Trish is currently teaching in the Masters of Education in the course Creating Classrooms of the Future, and co-coordinates and teaches an online course in the Graduate Certificate in Higher Education.

Angela D. Benson is an associate professor of educational technology at the University of Alabama, where she researches the influence of various uses of educational technology on individuals and organizations. Dr. Benson has designed and taught a variety of traditional and distance courses. She holds a doctorate in Instructional Technology from the University of Georgia. Angela D. Benson, University of Alabama.

Richard Caladine is Manager, Learning Facilities and Technologies at the University of Wollongong, Australia. He and his team supervise the teaching and learning facilities as well as the audiovisual, podcasting, videoconference, and video production services. As an academic Richard actively researches the use of rich media technologies in teaching and learning, with an emphasis on the higher education sector. The theories of technology selection he has developed and practical applications of them are described in his many publications. Richard's doctoral students are researching the use of rich media for teaching and learning between the University of Wollongong's nine campuses and Access Centres. Richard's current research area concerns the use of collaborations and unified communications in learning.

Brendan Calandra, PhD, is Associate Professor of Learning Technologies at Georgia State University. He is interested in the design of digitally enhanced learning, and applies his energy and expertise to teaching and conducting design-based research. His emphasis is on using emerging technologies for professional development. Some of his published work can be found in *Educational Technology Research and Development; Educational Technology; Journal of Research on Technology in Education; Journal of Technology and Teacher Education; Technology, Pedagogy, and Education;* and *Journal of Educational Multimedia and Hypermedia.*

Aleco Christakis has thirty-five years of experience in developing and testing methods for engaging stakeholders in productive and authentic dialogue for the practice of participative democracy. He is the author of more than a hundred papers on stakeholder participation, most recently the book *How People Harness their Collective Wisdom and Power to Create the Future* (2006). He is a co-founder of the Club of Rome, and past President (2002) of the International Society for the Systems Sciences (www.ISSS.org).

Alec Couros, PhD, is the Coordinator of Information and Communication Technology and a professor of educational technology and media at the Faculty of Education, University of Regina. Alec teaches and works with faculty, undergraduate, and graduate students to develop technology competencies and to promote pedagogical innovation. He also works as an administrator and research consultant with the Saskatchewan Instructional Development and Research Unit, where he leads research and training initiatives focused on the adoption of emerging technologies in teaching and learning. More information regarding Alec's teaching, research, presentations, and interests can be found at his blog, http://couros.ca.

Michael Dowdy has worked in the financial, healthcare, and defense industries over the past decade as a workforce learning professional. His primary interests involve the intersection of individualized informal learning and human capital systems within organizations. He has a Masters of Science degree from the University of Memphis with a focus in Instructional Design and Technology, and is currently working towards his Doctorate of Education.

Dr. Margaret Edwards, RN, PhD is a professor and Coordinator of Graduate Programs in the Centre for Nursing and Health Studies (CNHS) at Athabasca University, which offers the interdisciplinary Masters of Health Studies (MHS) and the Masters of Nursing (MN), Generalist and Advanced Nursing Practice. Her research interests are in the areas of healthcare informatics, online education, and exemplary care of the elderly.

B.J. Eib is an instructional designer with the Centre for Teaching and Educational Technologies, Royal Roads University. B.J. has a strong background in professional development with an emphasis on the effective use of educational technologies. She was a K–12 teacher for many years and then served as Associate Director of the Center for Excellence in Education at Indiana University. More recently, B.J. provided faculty development and served as the Inquiry and Blended Learning program consultant for the Teaching and Learning Centre at the University of Calgary.

Robert (Bob) Heller, PhD, obtained his doctorate in Experimental Psychology in 1992 at the University of Alberta and held a Post-Doctorate Research Fellowship from 1992 to 1994 in the Centre of Excellence on Aging Research Network where he conducted research on driving, dementia, and aging. He joined the Centre for Psychology at Athabasca University in 2001 and became interested in conversational agents and their role in distance education. This research has evolved into an investigation of animated historical figures as pedagogical agents and their place in immersive worlds.

Rita Kop has recently started work as a researcher for the National Research Council of Canada on their PLE project. Before that, she was an assistant professor in Adult Continuing Education at Swansea University in the UK, after a career in Dutch primary education. She has written extensively about widening access to higher education and e-learning. Her research interests are personalized learning; distance education and learning through online networks; the impact of technology on the concept of knowledge, learning and teaching; widening access to higher education; and institutional change.

Romina Laouri works for Ashoka, an organization that promotes social entrepreneurship around the world. She completed her undergraduate in Political Science, International Studies, and Spanish at Macalester College (Minnesota, U.S.), and an MA in International Relations at the University of Chicago. In Cyprus, she co-founded Youth Promoting Peace, a group that facilitated dialogue between young Greek- and Turkish-Cypriots, and she currently serves as Board Member to the Cyprus Neuroscience and Technology Institute.

Yiannis Laouris, neuroscientist and systems engineer, has applied linear/non-linear methods for over fifteen years to study the brain. In the 90s he founded Cyber Kids, a chain of computer learning centres, which pioneered the development of an innovative, socially responsible, educationally relevant method to introduce IT to almost 15 percent of the child population of Cyprus. He and his team have used the science of Structured Dialogic Design in more than thirty settings. He is currently the President of Cyprus Neuroscience Technology Institute.

The Learning Technologies Collaborative at the University of Minnesota represents the collective design, research, and integration work of Aaron Doering, Charles Miller, and Cassandra Scharber, in no particular authorship order. Our work is focused on creating innovative interaction design opportunities, researching contemporary pedagogical integration models, and developing practical frameworks of learning and teaching in the context of transformative online learning experiences for K–12 and adult education.

Mark J.W. Lee is an adjunct senior lecturer with the School of Education at Charles Sturt University, Australia's largest distance education provider, and an honorary research fellow with the School of Information Technology and Mathematical Sciences at the University of Ballarat. He is Chair of the New South Wales Chapter of the Institute of Electrical and Electronics Engineers (IEEE) Education Society, and serves on the editorial boards of several international journals in educational technology and e-learning.

Trey Martindale works with talented graduate students as Associate Professor of Instructional Design and Technology at the University of Memphis. He is a researcher with the Institute for Intelligent Systems, creating tutoring and instructional systems of the future. Dr. Martindale's expertise is in the design of online learning environments. His research efforts have been funded by the U.S. Department of Education, the Institute of Education Sciences, the U.S. Department of Defense, and Microsoft Corporation. For more information, please visit his website, http://teachable.org.

Mary R. McEwen is currently a doctoral student in Instructional Psychology and Technology. She has taught classes in mathematics and computer technology at the high school and college level. Mary has many years' experience as a software engineer, architect, and instructional designer, and has helped develop educational software for such companies as WICAT Systems, IBM EduQuest, The Learning Company, Waterford Institute, and NetSchools Corporation. Her current research interests include using web analytics to inform and improve instructional design, computer adaptive testing, and psychometrics.

Catherine McLoughlin is an associate professor with the School of Education at the Australian Catholic University, Canberra. She also serves as the coordinator of the Australian Capital Territory hub of the National Centre of Science, Information Technology, and Mathematics Education for Rural and Regional Australia (SiMERR). Catherine has been Editor of the *Australasian Journal of Educational Technology* since 2002, and serves as an editorial board member of several leading journals, including the *British Journal of Educational Technology*.

Hagit Meishar-Tal (PhD) teaches at the Open University of Israel (OUI) in the MA program in Education (Technology and Learning). She is also a member of Chais (Research Center for the Integration of Technology in Education), and a member of the Instructional Design Group at SHOHAM (Center for Technologies in Distance Learning). Hagit specializes in collaborative learning pedagogies and practices, and led the implementation of wikis in the OUI. Hagit holds a PhD from Haifa University. Her dissertation title: "The Internet and social space dynamics: Globalization, networking and virtualization."

Dr. Beth Perry, RN, PhD, is an associate professor in the Centre for Nursing and Health Studies at Athabasca University. Beth is a registered nurse and has her PhD in educational administration and policy studies from the University of Alberta. Beth's research interests are in exemplary online teaching and career satisfaction in nursing. Beth teaches in the undergraduate and graduate nursing and health studies programs, focusing on leadership courses.

SaraJoy Pond is currently a doctoral student in Brigham Young University's Instructional Psychology and Technology program. She has worked as a marketing, advertising, web design, and strategy consultant in the technology, entertainment, and services sectors, and as an evaluator in several nonprofit and educational contexts. Her research interests centre on capacity-building and technology integration in design and evaluation efforts towards sustainable development.

Mike Procter, P.Eng, obtained his BSc in Electrical Engineering in 1982 and is a registered Professional Engineer in Alberta, Canada. He has over twenty-five years' experience in the information technology industry, with a background in real-time systems development, system and network management, project management, and enterprise application deployment. He is currently an IT consultant working for Athabasca University developing software for research in animated conversational agents.

P. Clint Rogers holds a doctorate in Instructional Psychology and Technology, and is active in consulting, teaching, and research. He is currently adjunct faculty, teaching courses on web analytics, cross-cultural communication, and innovation. Clint is frequently requested as a consultant for international organizations regarding web analytics' application in business and education. Working with the University of Joensuu, Finland, he coordinates the ICT4D Consortium of African and European Universities and supervises dissertations in the IMPDET program (International Multidisciplinary PhD Studies in Educational Technology, www.impdet.org). To see his most recent research interests and discoveries, you can visit him online at http://www.clintrogersonline.com/blog.

Robyn Smyth is Senior Lecturer working as academic developer in the School of Rural Medicine at the University of New England, Australia, specializing in curriculum design, using rich media technology in higher and medical education and for student-centred pedagogies. She has been an active researcher investigating video communications for a number of years, principally focused on the potential of rich media technologies to enhance distance learning in higher education.

Currently Robyn is leading a sector-wide investigation into the sustainability, viability, and scalability of rich media technologies in Australian universities.

Edna Tal-Elhasid is the head of SHOHAM (Center for Technologies in Distance Education at the Open University of Israel). She is a member of Chais (Research Center for the Integration of Technology in Education) at the Open University of Israel. Her main areas of interest are open educational resources and open e-books, online collaborative learning and Web 2.0 for distance and higher education, social networks, and online teaching and learning. Most of her papers refer to the use of wikis at the Open University of Israel.

Belinda Tynan is Academic Director of the Faculty of The Professions at the University of New England in Australia. As Academic Director, she contributes to the development of academic policy within the university, with a particular focus on fostering strategic change. Belinda is also an editor of *Higher Education Research and Development*, a refereed international journal published by Taylor & Francis. Belinda is Treasurer of the Open and Distance Learning Association of Australia (ODLAA). Her research background is in the area of academic development, distance education, and models of research collaboration.

Gayle Underwood has consulted in schools all over Michigan and around the world, including Kuwait, where she worked with the staff at the Khalifa School in Kuwait City to help them implement Assistive Technology with their students. Recently, through her involvement with the Michigan's Integrated Technology Supports initiative at the Universal Design for Learning approach for Michigan's schools, her work expanded into learning the theory and practice of the science of Structured Dialogic Design.

Deborah Vale has worked in academic development, completing a number of small research projects for the Learning and Teaching Centre at the University of New England, Australia. She presently manages the

Rich Media Project investigating the use of rich media for improving student outcomes. Deb completed her doctorate at the University of New England in archaeology, specializing in marine faunal remains from coastal archaeological sites. She has worked as an archaeologist and palaeoichthyologist in Australia and Central America.

George Veletsianos, PhD, is Assistant Professor of Instructional Technology at the University of Texas, Austin. His research and teaching interests involve the design, development, and evaluation of digital learning environments, and his focus areas are adventure learning, virtual characters, and the learner experience. George holds MA and PhD degrees in learning technologies from the University of Minnesota, Twin Cities, and further information about him can be found at www.veletsianos.com.

Charles Xiaoxue Wang, PhD, is an assistant professor of instructional technology at Georgia State University in the U.S. He is the Asia and Pacific Region Coordinator for the International Division of the Association for Educational Communications and Technology (AECT) and the Director of Research and Public Relations for the Society of International Chinese in Educational Technology (SICET, 2007–2010). His research interests include online environments for learning and training, instructional technology consulting, and technology integration in school environments. Some of his publications can be found in *TechTrends, International Journal of Instructional Media, Journal of Educational Technology Systems*, and *International Journal of Technology in Teaching and Learning.*

Elizabeth Wellburn has been with Royal Roads University in various capacities since 1999 and is currently an instructional designer with the Centre for Teaching and Educational Technologies responsible for liaison with the School of Environment and Sustainability. Before coming to Royal Roads, she spent many years in positions where she was involved in the use of technology for learning, including corporate sector training and also within the provincial government's Ministry of Education.

Andrew Whitworth earned his doctorate in politics from the University of Leeds in 2001, with a study of communicative practices in environmental organizations, and he has since transferred his interest to the way ICTs affect communication, management, and professional practice in higher education. He is the course director of Manchester University's MA in Digital Technologies, Communication, and Education (http://www.MAdigitaltechnologies.com).

Professor Yoav Yair has been Director of the Center for Technology in Distance Education in the Open University of Israel since 2003. He is a physicist by training, specializing in planetary atmospheres and lightning. Professor Yair has a strong background in educational technology and science education, with specific focus on virtual reality and scientific visualizations. Since 2006 he has been leading the OCW project at the Open University, and directs the activities for planning the strategic usage of information technologies in higher education.

Youngjoo Yi, PhD is an assistant professor in ESOL/literacy in the College of Education at Georgia State University in the U.S. Her research interests include adolescent English language learners' multiple literacies (including online literacy practices) and their identity construction. Her work has been published in various journals, including *International Journal of English for Specific Purposes, Journal of Second Language Writing, Journal of Adolescent and Adult Literacy,* and *Journal of Asian Pacific Communication.* Recently, she guest-edited a special issue on "Biliterate Asian Students' Literacy Practices in North America" for the *Journal of Asian Pacific Communication.*

INDEX

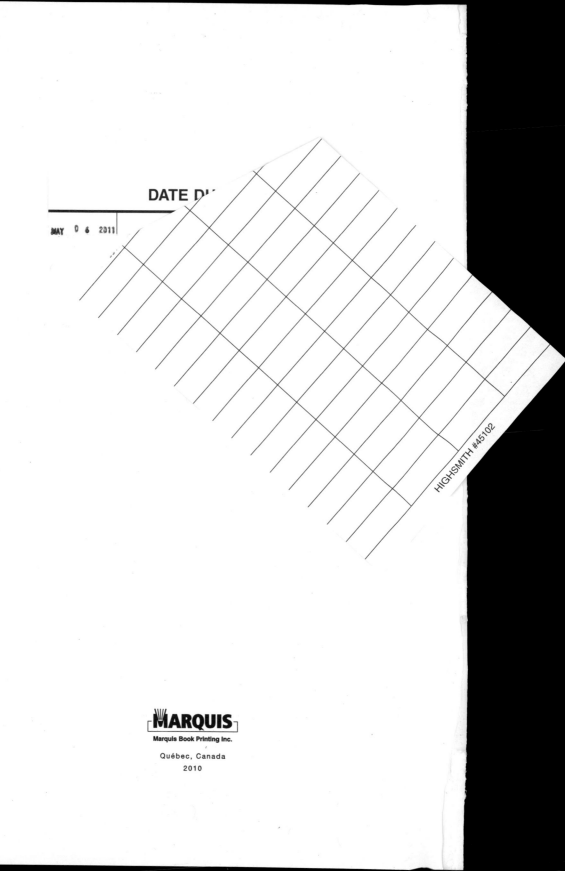

MARQUIS

Marquis Book Printing Inc.

Québec, Canada
2010